DARWINISM IN ARGENTINA

DARWINISM IN ARGENTINA

Major Texts, 1845–1909

Edited by
LEILA GÓMEZ

Translated by Nicholas Ford Callaway

Lewisburg
BUCKNELL UNIVERSITY PRESS

Published by Bucknell University Press
Co-published with The Rowman & Littlefield Publishing Group, Inc.
4501 Forbes Boulevard, Suite 200, Lanham, Maryland 20706
www.rowmanlittlefield.com

Estover Road, Plymouth PL6 7PY, United Kingdom

British Library Cataloguing in Publication Information Available

Library of Congress Cataloging-in-Publication Data

Piedra del escándalo. English
Darwinism in Argentina : major texts (1845-1909) / [edited by] Leila Gómez.
p. cm.
Includes bibliographical references.
ISBN 978-1-61148-386-4 (cloth : alk. paper)—ISBN 978-1-61148-387-1 (electronic)
1. Darwin, Charles, 1809-1882. On the origin of species. 2. Darwin, Charles, 1809-1882. 3. Evolution (Biology)—Argentina—History—19th century—Sources. 4. Evolution (Biology)—Social aspects—Argentina—History—19th century—Sources. 5. Argentina—Intellectual life—19th century—Sources. I. Gómez, Leila. II. Title.
QH365.O8P5413 2011
576.8'20982—dc23 2011026689

™ The paper used in this publication meets the minimum requirements of American National Standard for Information Sciences—Permanence of Paper for Printed Library Materials, ANSI/NISO Z39.48-1992.

Printed in the United States of America

CONTENTS

SOURCES AND ACKNOWLEDGMENTS

William Henry Hudson, the Woodpecker of the Pampas. Source: Hudson, William Henry. "Third letter," Proceedings of the Zoological Society of London 47 (1870): 158–706.

Charles Darwin, The Response. Source: Charles Darwin, "Notes on the habits of the pampas woodpecker (*Colaptes campestris*)," *Proceedings of the Zoological Society of London* no. 47 (1870): 705–6.

Francisco Javier Muñiz, The Muñi-Felis Bonaerensis. Source: EN: From the *Gaceta Mercantil*, 1 July 1845. Also found in Francisco Javier Muñiz, *Escritos Científicos* (Buenos Aires: W. M. Jackson, 1953), 20–44.

Charles Darwin, A Letter to Muñiz. Source: Charles Darwin, "A Letter to Muñiz," Nature 99 (1917): 305–6.

Domingo F. Sarmiento, Ñata Oxen. Source: Domingo F. Sarmiento, *Vida y escritos del coronel D. Francisco J. Muñiz* (Buenos Aires: Félix Lajouane: 1885), 257–267.

Eduardo L. Holmberg, Two Parties in Conflict. Source: Eduardo Holmberg, *Dos partidos en lucha* (Buenos Aires: Imprenta del Argentino, 1875), 89–108.

Eduardo L. Holmberg, Charles Robert Darwin. Source: Eduardo Holmberg, *Carlos Roberto Darwin* (Buenos Aires: Establecimiento Tipográfico de El Nacional, 1882).

Francisco P. Moreno, The Study of the South American Man. Source: Francisco P. Moreno, *Estudio del hombre Sud-Americano* (Buenos Aires: Imprenta La Nación, 1878).

Domingo F. Sarmiento, Darwin. Source: Domingo F. Sarmiento, *Obras*, vol. 22 (Buenos Aires: Imprenta Mariano Moreno, 1899), 104–133.

Juan Bautista Alberdi, Luz del Día's Pilgrimage. Source: Juan Bautista Alberdi, *Peregrinación de Luz del Día, ó viaje y aventura de la verdad en el Nuevo Mundo* (Buenos Aires: Casavalle, 1875).

Hermann Burmeister, History of Creation. Source: Germán Burmeister, *Historia de la Creación* (Madrid: Juan Ulled, 1900).

Florentino Ameghino, Remembrance in Honor of Darwin. Transformism as an Exact Science. Source: Florentino Ameghino, *Conceptos fundamentales* (Buenos Aires: W. M. Jackson, no date), 58–87.

Leopoldo Lugones, Tribute to Ameghino. Source: Leopoldo Lugones, *Elogio de Ameghino* [Tribute to Ameghino] (Buenos Aires: Otero, 1915), 37–64.

José Manuel Estrada, Naturalism and Education. Source: José Manuel Estrada, *Obras completas*, vol. 12, *Discursos* (Buenos Aires: Compañía Sudamericana de Billetes de Banco, 1905), 207–29.

Leopoldo Lugones, Izur. Source: Leopoldo Lugones, *Las fuerzas extrañas* [Strange Forces] (Buenos Aires: Arnoldo Moen y Hermano, 1906).

José María Ramos Mejía, The Argentine Masses. Source: José María Ramos Mejía, *Las multitudes argentinas* (Buenos Aires: Coni hnos, 1899), 294–321.

Carlos O. Bunge, Our America. Source: Carlos O Bunge, *Nuestra América: Ensayo de psicología social* (Buenos Aires: Casa Vaccaro, 1915 [1903]).

Jose Ingenieros, Sociological Evolution in Argentina: from Barbarism to Imperialism. Source: José Ingenieros, *La evolución sociológica argentina* (Buenos Aires: Librería Jiménez: 1910), 11–41.

Domingo F. Sarmiento, Racial Conflict and Harmony in the Americas. Source: Domingo F. Sarmiento, *Conflicto y armonías de las razas en América* (Buenos Aires: S. Ostwald: 1899), 53–68.

ACKNOWLEDGMENTS

The English translation of this book was made possible with the help of the Kayden Research Award, given by the University of Colorado at Boulder in 2010. I would like to thank the members of the Kayden committee, as well as the Department of Spanish and Portuguese at Boulder, for believing in the importance of this project. Thanks to my colleagues and friends who read the manuscript partially or totally: Mary Long, Andrés Prieto, Peter Elmore, and Dulce Aldama.

I would also like to thank the Argentine editor of the original Spanish version of this book, Gastón Gallo of Ediciones Simurg, for so kindly clearing the way for the publication of the English version. And thanks as well to Dr. Greg Clingham, who enthusiastically accepted the book's publication at Bucknell University Press.

Very special thanks to Nicholas Callaway, the translator, for his dedication, thoroughness, and intelligence in turning very complex and sometimes obscure texts into clear and open material for readers of different disciplines.

Thanks also to my family in Argentina and the United States for their companionship and support. Thanks in particular to my husband Javier and my children Faustino and Gabriela for always happily and generously sharing their time with my passion for this book.

ABBREVIATIONS

OAN: Original Author's Note
EN: Editor's Note
TN: Translator's Note

INTRODUCTION

Darwin, the Heart of the Controversy: Science, Politics, and Literature

Leila Gómez

This book documents the actual controversy underlying the fictional feud between Darwinists and anti-Darwinists in Eduardo Holmberg's novel, *Two Parties in Conflict* (*Dos partidos en lucha*, 1875). In the novel an elderly Darwin travels to Buenos Aires in order to demonstrate his theory of evolution to these two Argentine factions. In real life, Darwin traveled to Argentina only once, as a young man aboard the *Beagle*. It was only years later that his ideas came back to Argentina in the form of books and articles that would shake the foundations first of science, and then of sociopolitical thought in Argentina. We will examine the legacy of the feud that was first taken up by Bishop Samuel Wilberforce and Thomas Huxley in the Oxford Evolution Debate of 1860, following the publication of *On the Origin of Species* in 1859. This debate quickly spread to the peripheral regions of Western science and thought, and made Darwin into "the heart of the controversy" in Argentina, to use the words of President Domingo F. Sarmiento in his 1882 homage to Darwin, included in this anthology. Here the scientific debate on evolutionism was taken up by such figures as Sarmiento, Hermann Burmeister, Francisco P. Moreno, Florentino Ameghino, and Holmberg himself.

During his voyage aboard the *Beagle* from 1832 to 1836 through South America and the Pacific, Darwin began to make conjectures about and find evidence for the theory that would bring him both fame and invective. In Argentina he made observations about geology, and found answers to his questions in the pampas, which he referred to as "one wide sepulchre of extinct gigantic quadrupeds."[1] He collected information not only on the pampas's enormous fossils, but also on the area's stratigraphic sequences, and its modern-day species wild and domesticated, all of which would resurface in *On the Origin of Species*. Further along in his voyage, on the Galapagos Islands, he observed how the archipelago's multiple varieties of finch exhibited surprising adaptations to isolated environments, which led him to conceive of the changeability of species as the most plausible explanation for the creation of life on this planet. After returning to England, Darwin continued his research in conjunction with other scientists, such as the geologist Charles Lyell and the botanist Joseph Hooker. After more than twenty years of research, on the advice of his friends he finally decided to publish a portion of his theory before Alfred R. Wallace, who by a different route had arrived at similar conclusions concerning evolution, could beat him to it.[2]

The clergy waited until the third printing of *On the Origin of Species* to launch a public controversy, the fear of which had led Darwin to wait so many years before publishing his ideas. The creation of the world in seven days was a deeply rooted belief in Victorian England, despite the fact that scientists like Charles's grandfather Erasmus Darwin, or Georges L. Leclerc Comte de Buffon and Jean B. Lamarck, had already held an intuitive belief in evolution. Darwin was aware of this, and understood the enormous political and epistemological challenges that his findings would pose.

The English clergy could not stand idly by while the Bible's Week of Creation and the myth of Adam and Eve were being questioned. Wilberforce had announced that he was prepared to "tear Darwin apart" in the Oxford Debate, and he was supported by Richard Owen, famous for his outspoken rejection of natural selection.[3] Darwin, who for health reasons was unable to attend the debate, was nevertheless well represented and enthusiastically defended by T. H. Huxley and Hooker (Moorehead 119). Amid sarcastic remarks, insults, long speeches by scientific and ecclesiastical authorities, applauses and cries from the audience, Wilberforce asked Huxley if he preferred to claim descent from monkeys through his grandmother or grandfather, to which the scientist replied that he preferred to claim descent from a monkey than from a cultivated man who was opposed to the progress of science. This caused such an uproar of applause and shouting that several women began to faint and had to be escorted off the premises.

In Holmberg's *Two Parties in Conflict*, another debate takes place, here between the Rabianists or (anti-Darwinists) and the transformists (or Darwinists) in Argentina. The characters in each of the factions, although not completely transparent or readily identifiable, are associated with the historical figures of the anti-Darwinists Burmeister and Moreno, and the Darwinists Ameghino and Holmberg. Politicians like Sarmiento, Mitre, Alsina, and Avellaneda also participate. Furthermore, each faction represents a position which is not only scientific but also political and philosophical. The monologue of Dr. Griffritz, the most prominent of the Darwinists, sums up the ideology of each group in this intellectual climate:

> If the Rabianists triumph we shall see the proliferation of the status quo, with all of its excess, with all of its fatuous stability. The sciences will not advance, and even if they do it will be a negative advancement, as I see it. If on the other hand we the Darwinists triumph [. . .] it goes without question that the entire social norm will have to change, and, either a philosophical revolution of incalculable transcendence will erupt, or an indifference will come about so extreme that we will no longer be able to appreciate the influence of a scientific doctrine on the progress of society.

The historian Marcelo Montserrat, in his exhaustive study, has called Darwinism the "ideology of progress," and points out that at the turn of the century in Argentina's elite circles, both Darwinism and the positivism of the preceding decades were bywords for the modernization of political and social institutions, and the secularization of the state. Progress was understood in terms of evolution and the overcoming of the "primitive stages of mankind," an idea which fit hand in glove with Argentina's wars against outlying and indigenous populations, as in the Desert Campaign.

At the same time, Holmberg's novel presents a "scathing picture of the growing urban demonstrations" and the way in which crowds, much like the one at Oxford, intervened in the political life of fin de siècle Buenos Aires.[4] Though in the novel common people are merely used as a passive audience for the spectacle of science and politics, in later decades these "masses" would become the very object of the medical and social-Darwinist analyses of José M. Ramos Mejía, Carlos O. Bunge, and in some ways José Ingenieros as well, all of whom are represented in this anthology.

Inspired by the controversy between creationists and evolutionists, this anthology studies in depth the debate in Buenos Aires between Holmberg's Rabianists and transformists, and reconstructs more than a half century of the dialogue and philosophical, political, and scientific altercations surrounding Darwinism in the Argentine intellectual context. This book begins with the first readers of Darwin in Argentina: William Henry Hudson, Francisco Javier Muñiz, and Domingo F. Sarmiento. It then brings together texts from the period in which the sciences were institutionalized in Argentina, with the creation of museums, scientific societies, and participation in fairs and worldwide exhibitions. The authors from this period are Hermann Burmeister, Florentino Ameghino, Francisco P. Moreno, Eduardo L. Holmberg, and the Catholic José Manuel Estrada. In the last section, we will turn to the period of the centennial of Argentina's independence, wherein the positivist and

Darwinist school exhibit a strong "tendency toward dogmatism," as in the racialist proposals of Ramos Mejía, and Bunge.

In these medical-psychiatric texts, Argentine lawyers and sociologists propose an evolutionist reading of Argentina's history, and portray their society as a "sick body." In this interpretation, the intellectual elites apply the Darwinist concepts of involution and atavism to explain what they see as the immigrant's "psychological pathologies."

Darwin is one of a number of English travelers who first described the natural world of Argentina after its independence from the Spanish crown.[5] Roberto González Echevarría claims Latin American national discourses following independence appropriated the genre of the scientific travel narrative to distance themselves from the colonial bureaucracy and legal discourse of the Spanish Empire.[6] The trait that defines "Latin American Otherness" is the vastness of its wilds. For González Echevarría,

> the Latin American self both fears and desires that Other Within, because of his lawlessness, and travels to meet him. But the only way to apprehend him is through the mediation of a hegemonic discourse, which is now that of modern science, as disseminated by the naturalists who turned the New World into their living laboratory.[7]

Thus, European science became a more "modern" and appropriate discourse after the independence of the new republic. This anthology brings together texts from Argentina that sought to appropriate Darwin as a scientist-traveler in hopes that his theoretical perspective would explain that "Otherness" and help them with the task of building a nation.

What is certain is that Darwin has aroused and continues to arouse feelings of national pride in Argentina. Sarmiento used to say that the Argentinians "have given . . . science and fame to Darwin, with Argentine fossils and stock." José Babini recalls that Sarmiento used to jokingly boast that the theory of evolution was Argentinian, given that during his trip through Patagonia, Santa Fe, and the Andes, Darwin was able to observe and gather the physical evidence for the theory that he would go on to develop.[8] On the other hand, Aníbal Ford is careful to point out that the scientific aspects of the Beagle's voyage—whose main attraction was cartography for Fitz-Roy and naturalism for Darwin—overlapped with the political and economic interests of Great Britain in the Falkland Islands which, since the late 1820s, English sailors and merchants had seen as a secure port for their voyages to Australia and Tasmania.[9] For Ford, the voyage of the *Beagle* was not simply motivated by "the most noble purpose of science,"[10] but served imperialist objectives as well. From the point of view of Mary Louise Pratt, the scientist-traveler's obsession with distancing himself from political and economic missions stands in contrast to the overall enterprise of his voyage. For Pratt as well as Ford, the metropolitan scientist-traveler looks to secure "his innocence" or disinterest in relation to imperial missions, but nevertheless contributes more or less consciously—with his measurements; preparation of maps; and physical, anthropological, and geographical descriptions—to securing a clear field of vision for the imperial gaze in the explored territories.[11]

PART I: THE FIRST READERS OF *ON THE ORIGIN OF SPECIES* IN ARGENTINA: THE DIALOGUE BETWEEN DARWIN AND MUÑIZ AND THE CONTROVERSY WITH HUDSON

In his book *Science in Action*, Bruno Latour claims that science's basic task is the progressive accumulation and mobilization of distant events, places, objects, and people. Accumulation would be carried out in the metropolis, understood to be the "center of calculation,"[12] and its institutions would determine how this accumulation and displacement of objects will be read. The use of the term "accumulation," with its economic implications, is not arbitrary, for the networks of scientific exploration correspond with Western networks for the exchange and accumulation of capital. Latour argues that the asymmetric relationship between knowledge in the metropolis with respect to the colonies was made possible by the former's ability to position itself at the center of the network, and act from a distance on colonial events, places, objects, and peoples, so as to bring them "back home." To achieve this goal, the proper procedure is to cut out and construct objects that are "mobile and stable," so that they can be displaced in both directions without being distorted by their decontextualization. These mobile and stable objects should therefore be made "combinable" and "theorizable" with other distant objects in museums, libraries, universities, and scientific societies, i.e., the institutions at the "center of calculation." A fairly clear example of this cycle of accumulation is the voyage of Darwin aboard the *Beagle* and the collections of specimens and fossils from the pampas and Patagonia that he amassed, as well as their subsequent theoretical explanation and dissemination through publications and academic circles across England and Europe. According to Latour's model, the New World played a big role in Darwin's theory.[13]

Local naturalists like Hudson (1841–1922) and Muñiz (1809–1882) collaborated in this cycle of accumulating and producing scientific knowledge for the metropolis. At twenty-four Hudson began a scientific exchange with the Smithsonian Institute in Washington, specifically with the ornithologists Spencer Fullerton Baird and Phillipe Lutley Scatler. This exchange was facilitated by Hermann Burmeister, then the director of the Museum of Natural Sciences of Buenos Aires. From that time on Hudson sent them embalmed specimens of native species and, in doing so, he became the intermediary in the kind of mobilization of scientific capital that Latour describes. The metropolis named at least two of the bird species sent by Hudson in his honor: *Knipolegus hudsoni* (Hudson's black-tyrant) and *Asthenes hudsoni* (Hudson's canastero). From 1869 to 1870 Hudson sent not only stuffed birds but also his writings on avian behavior. These writings were edited by Scatler and published in the *Proceedings of the Zoological Society of London*. In this way, metropolitan science had access to Hudson's "firsthand" knowledge, even before the local knowledge networks had the resources necessary to disseminate this information within Argentina's scientific circles. A few years later, in 1873, the mobilization was complete, and Hudson the naturalist moved to London permanently.

This anthology includes one of these pieces written by Hudson for the *Proceedings*. It is part of an exchange between Hudson and Darwin concerning the pampas woodpecker. Marcelo Montserrat has pointed out that Hudson was really the first reader of Darwin in Argentina. He was a reader with a unique linguistic background, born into a family from the United States that had moved to the pampas, where he was raised among Irish tutors and English gauchos. This background gave Hudson early access to *On the Origin of Species*: on a trip to England, Hudson's older brother obtained a copy for him, which he read avidly as an adolescent. After this reading, the spiritualist and animist Hudson asked himself: "How was I to reconcile the ancient fables and notions [of my mother's transcendentalism] with the doctrine of evolution? What effect had Darwin had on me?"[14] Hudson is, then, one of the first to juxtapose, though not combine, Darwinism and animist spirituality. A few decades later, spirituality and science would fuse in Holmberg's *La bolsa de huesos* (*The Bag of Bones*, 1896), in Lugones's *Fuerzas extrañas* (*Strange Forces*, 1906), and also, to a certain degree, in Ameghino's *Mi credo* (*My Creed*, 1906).

"The Pampas Woodpecker," is one of the most important pieces of those compiled by Scatler in the aforementioned London periodical. On this occasion Darwin himself rebuts Hudson's objections, on both scientific and moral terms. While he is but a gaucho in the pampas, Hudson thus has access to Darwin's scientific circle through Scatler. Although Hudson makes some strong allegations against Darwin, the controversy between the metropolitan and the local never transcended the epistolary debate. If read carefully, Hudson's observations do not, in fact, refute Darwin's theory, but rather constitute his own minor contribution to it.

Based on his observations of the pampas woodpecker, Darwin had concluded that it constituted the strangest case of this family he had found during his travels aboard the *Beagle*, given that the bird's traits, despite the fact that it lived in areas devoid of trees, had not been modified.

As Hudson sees it, Darwin's conclusions rest upon observations made over the course of just a few days' travel through the area, the sort of hasty formulations that would serve as fodder for the theory of evolution's detractors. To Darwin's claim, Hudson counters that the pampas do indeed boast the *ombú* tree, and in areas not far from where Darwin made his observations. This, he asserts, is the explanation behind the unmodified woodpecker. What is interesting about this exchange is that Hudson uses his local experience as symbolic capital to legitimate himself within the metropolitan scientific community that has taken him in through the *Proceedings*. In this way, Husdon positions himself as the best authority to describe the birds that live in the pampas. Unlike Darwin's sporadic observations, Hudson's claims are the product of regular contact with the region's fauna, and are largely based on Félix de Azara's authoritative study.

The clash that Hudson taps into between local and metropolitan is not new; what may be new is that Darwin decides to take part in the debate and, in doing so, modifies the status of Hudson's accusations. In Darwin's reply, Hudson's "local" knowledge becomes "scientific" knowledge that is worthy of a response, by virtue of the complex mechanisms that authorize cycles of scientific accumulation and institutions of power.[15] Even still, years later, an elderly Hudson living in England would express his regret for what he came to see as youthful effrontery.

Francisco Javier Muñiz also carried on a correspondence with Darwin, though in this case a private one. Included in this anthology is the letter that Darwin sent Muñiz thanking him for his descriptions of the *ñata* cow and the Muñi-Felis fossil, and for offering to sell his fossil collections. When Darwin traveled through Argentina he did not meet this Luján's resident soldier, doctor, and collector of fossils. However, the observations on the *ñata* cow that Muñiz sent him proved crucial to the theory of evolution and to Darwin's studies into the mutations of domesticated animals. For Julio Orioni and Fernando Rocchi, Muñiz begins the series of fruitful exchanges between Darwinism and Argentine science. Although he never called himself a Darwinist, it was Muñiz who transmitted his fervor and enthusiasm about evolution to Sarmiento.[16]

Both Sarmiento and Ameghino studied Muñiz's biography and writings in depth. Both of them recognized his contributions as a paleontologist; as a military doctor in Cepeda and in the Paraguay war; and as a university professor and chair of the department of "theory and practice of birth, childhood diseases, and medical jurisprudence"; as well as his work on health reforms in Argentina. One of the highlights of the excavations he carried out during his fifteen-year residency in Luján was the discovery of the first complete *Megatherium*, which was sent to the Cabinet of Natural History in Madrid. His numerous finds enriched collections near and far, in Sweden, France, and Buenos Aires, as a significant portion of the extremely valuable fossils that Muñiz offered to President Rosas, in hopes of founding a natural history museum, were to end up in foreign hands.

Sarmiento points out that "with Muñiz there began, in this country, a scientific and literary movement aimed at studying ourselves and the country in which we live."[17] It could be said, then, that Muñiz actively participated in the mobilization of fossils through the cycles of scientific accumulation, facilitating their transfer to the metropolis. However, at the same time, he participated in a reverse cycle of accumulation, which would eventually lead to the preservation of the nation's heritage, and its capitalization and exhibition in local museums and institutions.

Muñiz's study of the Muñi-Felis fossil included in this anthology is important because, aside from having been a topic of conversation with Darwin, it triggered a debate over the size of American fauna. Buffon had claimed that both the fauna and the people of the New World were inferior in size to those of the Old World. He and others saw a clear parallel between the size of the "primitive" regions' species on the one hand, and their supposedly childlike character on the other. It was not only the natural sciences that adopted this incipient evolutionist model; it was utilized by philology as well in its classification of "primitive," "barbarian," or "savage" languages, as opposed to the "civilized" ones, that is to say, those that were Indo-European, Arabian, or Chinese in origin. Thus, in the same way that the New World puma was smaller than the Old World lion, the Amerindian languages were wrongly perceived to be homologously childlike, due to their supposed inability to express universal or abstract thoughts, and moreover for their alleged lack of any non-iconic writing systems or sophisticated technology, such as alphabetical writing.[18]

What the Muñi-Felis demonstrated—according to Muñiz—was that there had once been a big cat whose size was greater than that of Old World felines, meaning that the fossil flew in the face of the entire system of prejudices and beliefs that had devalued New World wildlife. Muñiz takes it upon himself to correct this prejudice of Buffon, exhibiting a criollo's[19] drive to redeem the natural world of the Americas in the eyes of the emerging field of paleontology.

In studying the writings of Muñiz, Sarmiento pauses to reflect on the description of the *ñata* cow that Muñiz sent to Darwin. Making use of both Muñiz and Darwin's studies of this "indigenous" variety of pampas stock, Sarmiento speaks of the need to "civilize" the cattle. By this he means that the Argentine countryside should be modernized, so that Argentina might better position its products in the world market. The *ñata* cow, Sarmiento points out in the article included in this anthology, is of particular interest to Darwin, for it is a domesticated animal which at the same time is an example of "Darwinian atavism," through which a domesticated variety regains lost ancestral traits as a result of human neglect. This means that the *ñata* cow is an example of "involution," which Sarmiento blames on the natives' indolence. This is why the *ñata* cow, though of significant scientific value, has little value at all on the international market. Sarmiento transforms a scientific theory into a matter of political economy and begins to foresee its possible applications.

This is no small matter in a country of stockbreeders. While on the one hand Sarmiento uses Darwin's writings to underscore the backwardness of the local *ñata* cow, in his speech in honor of the deceased naturalist he claims that "[i]ntelligent breeders of sheep are perfect and unrivaled Darwinists in the art of *varying species*. It was from them that Darwin took his first notions, right here, in our very pastures." It is true that techniques of artificial selection were not unknown to local stockbreeders, and that their products influenced Darwin's ideas about natural selection. In 1813 a British stockbreeder by the name of Hasley introduced Merino sheep into Argentina, and a few years later, in the 1820s, President Rivadavia imported Merinos along with South Down sheep, which were crossed with the local breeds in order to improve their quality. The year 1820 was also the year the *toro tarquinus*, initiator of generations of Shorthorn cattle, was brought to Argentina. The first records detailing the pedigree of the Shorthorns mark the beginning of scientific stockbreeding in Argentina. This was later followed by books such as Daniel Pérez-Mendoza's *Tratado del ganado lanar* (*Treatise on Wool-Bearing Stock*, 1858), or, later still, Juan Harrat's *Estudios prácticos sobre la cría y el refinamiento del ganado lanar* (*Practical Studies into the Breeding and Refinement of Wool-Bearing Stock*, 1885).

PART II: SCIENTIFIC FANTASIES: DARWIN'S FOLLOWERS AND DETRACTORS IN ARGENTINA (SARMIENTO, AMEGHINO, MORENO, HOLMBERG, BURMEISTER, LUGONES, ESTRADA, ALBERDI)

The authors in this second part were chosen because they all took part in one collective argument concerning Darwinism, the argument that inspired Holmberg's *Two Parties in Conflict.* The choice of texts in this section goes further than the controversy parodied by Holmberg, and reveals a whole array of underlying political, ethical, and economic questions that were affected by the idea of Darwinism in this period. This is to say, evolution and the idea of the struggle for life provided arguments that would be legitimizing principles in the military campaigns of annihilation and inclusion/exclusion of indigenous peoples in/from the modern nation. From two opposing points of view, Holmberg (1852–1915) and Moreno (1852–1911) debate the ethicality of the extermination of the indigenous peoples, although both seem to agree that their disappearance will be inevitable as evolution takes its natural course. Both men project onto the Argentine pampas and Patagonia the ideals of modernity and progress: order, industry, and the total integration of their country into the world commerce market. In "Charles Robert Darwin," a talk (part of which has been translated for this anthology) given in 1882 following Darwin's death, at the Teatro Nacional in Buenos Aires at the invitation of the Argentine Medical Circle, Holmberg doesn't shy from referring to this topic explicitly:

> Is the Indian's cause just?
>
> Arguing without too much dialectic, the Indian defends his land, which we have stolen from him, he wounds us, he kills us, he robs us.
>
> Does he do right? It is clear, or not. He fights for his life. But since the laws of nature work most noticeably, in their greater manifestations, upon larger groups of humans, rather than on individuals, all the opinions of Providence, justice, equity, brotherhood, which are nothing more than opinions separately rooted in each of us, fall apart in the presence of that common manifestation which is, in a way, the law of nature;—and we Whites, the civilized ones, the Christian ones, the ones armed with Remingtons, do away with the Indians, because the Law of Malthus stands above all of these individual opinions, which may be truly excellent, but which, either because humanity is still far from being civilized, or on account of any other cause, don't flesh out—and so, as we too fight for our lives, with good ideas, with good weapons, with good resources, we do nothing more than put into practice our upper hand.
>
> —"Do we do right?" This is a question.

> —"We fight for our lives." This is an answer.
> Reason, lastly, is a victory of organic progress. But victory, in any form, is a *reason* that overcomes all progresses.

The rational exploitation of the land—i.e., measuring it, delimiting it, calculating it, and turning it into commercializable and exportable goods—is the task envisioned by the modern nation. As a part of this modernizing impulse, and perhaps as a counterweight to it as well, the process of institutionalizing the national heritage was sped up, along with its study, exhibition, and display in museums, scientific observatories, and university departments. As far as the issue of the indigenous people is concerned, the extermination campaigns took place even as the state made a symbolic space for them in the national museum, and at the same time as anthropology, natural history, and archeology were being made into official academic disciplines. Paleontology placed the nation's explored territories and its territories yet to be explored within a much wider timeframe. Geological eras, antediluvian fossils, and primitive peoples situated a whole swath of the national territory in human prehistory. Moreno, in his article included in this anthology, thought that Argentina's paleontology should reveal to humanity "the secret of [its] first homeland [. . .] the precise location of the encampments of [its] ancestors, who with time would go on to build entire nations."

Thus, the evolutionary measure of modernity was progress, and that of the nation, its historical heritage. The unilinear frameworks drawn up by the nineteenth-century evolutionists were, just like those of the Enlightenment, logical constructs concerning human development through time. These frameworks consisted of a succession of uninterrupted states with no gaps between them, wherein the scientist's task was to go about ranking the world's human societies in accordance with their level of development along this scale. In this way both the contemporary Tasmanians and the primitive pre-Columbians of the Americas could be situated together at the beginning of this series of evolutionary states.

The theory of cultural evolution rested on the principle that cultural change is progressive, slow, natural, and inevitable.[20] However, cultural evolution proved to be fundamentally ethnocentric insomuch as it assumed that all cultures—past and present—represented milestones in the cultural development of the West, having advanced only part of the way before reaching a point of stagnation. For Montesquieu as well as Aristotle, Western culture had progressed along a set course, from most simple to most complex and from the lowest level, that of the savage, until reaching the highest level, that of civilization.[21]

In this manner, Sarmiento (1811–1888) says in "Darwin" that evolutionism—which he understands more in the vein of Spencer, i.e., social Darwinism—has given him "a principle that is both beautiful and harmonious, in order to dispel doubt, which is the torment of the soul." And so evolutionism would fill the void left after divine creation had been discredited, and would explain historical change, which Sarmiento conceived in terms of a progress that began its course in Classical Greece, was revived in the Renaissance with the invention of gunpowder and printing, with Copernicus and Galileo, passed through the Protestant reform, Bacon, Benjamin Franklin, and on down to Darwin.

The idea of social evolution underlies Sarmiento's belief in the civilizing power of education, in virtue of which a "barbarian" population—be it native or immigrant—can be adapted to the life of a modern and enlightened society. The selection included in this anthology from Juan Bautista Alberdi's book, *Luz del Día's Pilgrimage*, is an attack on this belief in the power of education to transform the species *Homo sapiens*. In this satirical tale we see how a Darwinist Don Quixote (Sarmiento) sets out to "evolve" a multitude of sheep (the masses) into human citizens, a scathing portrayal of Sarmiento's application of Lamarckian theories of moral inheritance to the realm of education. In Alberdi's (1810–1884) view, no amount of education can "improve" the human races; the South American populace, be it native or Mediterranean in origin, will never overcome its natural limitations (just as sheep cannot "be evolved" into regular human citizens). As he sees it, the fittest of the human races is the Northern European variety, an idea which was to hold sway in the country's ensuing debates over the best populations to target for immigration to Argentina.[22]

In "Mundos prehistóricos: Viaje aéreo a través del museo prehistórico de Moreno" ("Prehistoric Worlds: An aerial trip through Moreno's prehistory museum"),[23] Sarmiento—faced with the fossil material found by Moreno and evidence for different geological eras—agrees with contemporary naturalists about the possibility of setting the first chapter in the history of human evolution on Argentine soil. These "prehistoric worlds" will become the nation's heritage, its symbolic and religious capital. Heritage connotes ancestors, history, a shared past, and so it is based on these connotations that the discourse of power quickly incorporates heritage into its definitions of national identity, transforming it into that which belongs to "everyone" and to "no one," and establishes it as a tie of citizenship. In this way national heritage takes on the characteristics of what is "specific to our nation."

The fossil remains from the pampas and Patagonia in Argentina, together with their Tehuelche, Araucano, Mapuche, and Gennkene inhabitants, became museum pieces, which can easily be made to embody chronologies and national grammars. Beatriz González-Stephan has pointed out how, for outlying countries striving to be desired and consumed by the capitalist powers of the late nineteenth century, in fairs and museums, objects from the past acquired both aesthetic and symbolic value as pieces in collections and exhibitions.[24] The culture of exhibition and chronology is manifested in literary histories, museums, statues of epochal heroes, among whom scientists began to be included. Exhibitions and fairs were spaces in which science—as well as national heritage—proved instrumental to national legitimation. This is, for example, how Argentina experienced the recognition and gold medals awarded to Ameghino, Burmeister, the National Library, and the University of Buenos Aires, among other Argentine institutions, at the Paris World's Fair in 1889.[25]

The movement to institutionalize Argentina's sciences was charged with the task of creating and preserving the nation's heritage. One of the most prominent figures in this movement was the German national Hermann Burmeister (1807–1892), who was invited by Sarmiento to set in motion and give worldwide status to the nation's sciences, in particular the natural science museum. The museum was founded in 1812 by Bernardino Rivadavia, who, as a member and ideologue of the Generation of 1837, envisioned it as an icon of the new republic's autonomy. The foundation of the museum was linked with the birth of a sovereign and cultivated nation. This same liberal and enlightened spirit was to guide Sarmiento decades later.

Burmeister's arrival in Argentina was part of the accelerated institutional activity in the sciences that Argentina underwent in the seventies. The German national arrived in 1873, originally to take charge of the Faculty of Physio-Mathematical Sciences at the University of Córdoba. Two years prior the National Observatory had been inaugurated in the same city by Benjamin Gould. Also, in July of 1872 the Argentine Scientific Society was created, and in 1874 the University of Buenos Aires was recognized as such and began to admit students, among them Eduardo Holmberg and José María Ramos Mejía. During the same period Moreno and Ameghino had already begun their travels through the pampas and Patagonia.[26] In 1877 the Argentine Scientific Society, made up by Estanislao Zeballos, Valentín Balbín, and Miguel Puiggarí, among others, offered Darwin an honorary membership in the Society. By accepting, Darwin became the third major personality to receive this honor, after Guillermo Rawson, who had been a member since 1874, and Hermann Burmeister, since 1875.[27] Marcelo Montserrat notes that the anti-Darwinist Burmeister felt so irritated by this new addition to the Scientific Society that he tried, unsuccessfully, to change the statute for electing honorary members.[28]

Holmberg portrays the controversy between Burmeister and Darwin as a question of "amour propre." Holmberg quipped that among scholars, to say that one does not agree with someone else is like someone giving himself "a pat on the back, which posterity then uses to make for him a crown of laurels" (*Two Parties in Conflict*, 10). In his lecture "Charles Robert Darwin," he suggests that Burmeister was opposed to Darwinism "for the only plausible reason, which is that it is not called 'Burmeisterism'" (105).

Already in his *History of Creation* (1843),[29] the work with which he had gained an international reputation before arriving in Argentina (an excerpt of which is included in this anthology), Burmeister declared his opposition to Darwinism and stated that he did not believe in its "hypothetical conceptions without demonstrative power." For Burmeister, to accept Darwinism meant to open the flood gates of an epistemological revolution, to bring about a paradigm shift in the sciences over which, up to that time, he had held considerable authority:

> Man and monkey are presently distinguished zoologically and psychologically, and since we cannot simply drop the principle of the invariability of specific characteristics without at the same time revolutionizing the whole of scientific zoology, we are absolutely right in believing that their differences have existed since primitive times and indeed forever, and that they will continue to exist into the future.[30]

History of Creation espouses the beliefs that all the species are fixed, and that they are all imperfect realizations of one original ideal, an ontology that stands in clear opposition to Darwin's idea of the species' random variability. Burmeister held that this conception of fixed species modeled after one ideal was especially true of the genus *Homo*. In his theory there was an underlying, perfect human ideal that was altered by the influence of the different habitats in which each race was conceived. Thus, while some contend that Burmeister is a racist, others hold that, despite his opinion that there are innate differences between the races, the theory that they are all imperfect copies of one perfect ideal implies that no race is better than any other. His idealism in this respect contrasts with his strict empiricism in rejecting the theory of evolution for its lack of empirical evidence. For Burmeister, evolution was "outside the realm of just experimentation." In any case, the same can just as easily be said of his fixed ideal.[31] Moreover, current studies on genetics have proven Darwin right and Burmeister, wrong.[32]

In Argentina, Burmeister astutely carried out his official duties, arranging and disseminating worldwide the museum's collections, as well as taking a leading role in Argentina's scientific community of the period. Although he was opposed to transformism, through his travels and his descriptions of fossils Burmeister contributed to the accumulation of evidence in its favor. His study of horse fossils in the pampas, which he published in a bilingual German-Spanish edition for the Philadelphia Exposition of 1876, is an example of the sustained dialogue between Burmeister and Darwinism. Part of the fossil had been collected by Darwin during his voyage aboard the *Beagle*, and later on Muñiz was able to complete it thanks to further excavations. But it was Burmeister who brought order to the findings, describing the complete prehistoric horse fossil and comparing it to the modern-day horse. These descriptions questioned common beliefs about the American continents' native fauna.

The directorship of the Museum of Buenos Aires was also sought after by Ameghino and Moreno, among others, but the authoritative presence of Burmeister blocked these scientists from holding this lofty post. Nevertheless, thanks to the federalization of Buenos Aires and the importance that La Plata gained as capital of the province, the Museum of Natural Science of La Plata was founded. In this institution Moreno and Ameghino were able take the reins of the museum, jointly at the outset. This new institutional space was the result of the rivalry between the centralism of the city of Buenos Aires and the aspirations of the remainder of the province, represented since 1878 by Governor Carlos Tejedor. At the same time, the Museum of La Plata marked the opening of an alternative arena to that of Burmeister.[33]

The Museum of La Plata was created from Moreno's private collections, gathered over the course of ten years. Moreno's own prized possessions were transformed into pieces to be preserved and exhibited publicly, precisely when their scarcity or rarity—or "vanishing cultural status"[34]—had turned them into objects of value on the market of symbolic goods. Moreno pieced together these cultural assets during his trips to Patagonia before and after the military campaign decimated the region's indigenous inhabitants. His letters and travel diaries, as well as his *Reminiscencias* (*Reminiscences*), tell the story of a man who believed he was carrying out a messianic and foundational task. Not only did he found museums and set territorial boundaries between Argentina and Chile from 1897 to 1902; he also lived and negotiated with the native populations, and was even kidnapped on one occasion by the Mapuche chief Sahihueque in the vicinity of Lake Nahuel Huapi. Moreno represents the archetypal modern traveler who maintains an ambivalent position toward past and present, toward native populations and scientific progress. He was also able to maintain this ambivalence when faced with Darwinism, owing to the influence of his first mentor, Burmeister, even though—as Holmberg points out—with time he would become a "furious transformist" (*Charles* 106). As an evolutionist traveler he projected chronologies upon the landscape, the prehistoric animals, and the indigenous peoples of Argentina's national territory, describing them, drawing maps, and establishing the countries' boundaries in both space and time. Moreno felt he held a privileged position as he sutured and linked together the nation's past and present, endowing it thus with an order worthy of both science and spectacle.

The scientific conquest of Patagonia differs from the military conquest carried out by statesmen like Alsina and Roca. Nevertheless, Moreno's discourse also exhibits ambivalence in this respect: while on the one hand he openly criticizes the military campaign against the indigenous peoples, on the other his writings project upon the landscape a future modern and rational exploitation of the territory. His position is similarly ambivalent as far as the indigenous peoples themselves are concerned. For Moreno they are objects to study and to measure, evidence of past evolutionary periods naturally doomed to go extinct in the struggle for survival. However, his writings abound with compassionate assessments of the "primitive" life.[35]

Both Moreno and Ameghino (1854–1911) form part of the Argentine national pantheon. In his book, *Los arquetipos* (1922), Ricardo Rojas describes and glorifies what he considers to be the paradigmatic figures of the liberal enterprise, among whom Ameghino is to be found. Rojas chooses archetypal figures who embody the ideals of Argentine civilization, progress, and socialism. The greats of Argentina's history "parade" through the book's pages, from Manuel Belgrano, the epic hero of the country's independence, to figures like Sarmiento the educator, Carlos Pellegrini the statesman, Guido Spano the poet, and Florentino Ameghino. Rojas calls Ameghino "el sabio," meaning "the scholar" or "the wise man," and in referring to him and others like him he says:

> Heroes they were, and heroes struck in the purest of classical molds, the champions of emancipation who founded this nation on our native plot of land. Those champions tore asunder all forms of political injustice; but in spending Homeric days of toil measuring the homeland, and setting as its limits the mountain tops and the rivers which would redden its blood,

> they not only handed over the emancipated land to the joys of liberty, but also to all the responsibilities of civilization.[36]

Lugones wrote (1874–1938) *Tribute to Ameghino* (1915), which is at once a biography of the scientist and an explanatory treatise on his theories. Rojas and Lugones ushered in a trend of writing epic biographies that portray the nation's scientists as dignified role models, for example: *Ameghino (una vida heroica)*[37] (*Ameghino (A Heroic Life)*, 1951) by Fernando Márquez Miranda, *El perito Moreno: Centinela de la Patagonia* (*The Expert Moreno: Sentinel of Patagonia*, 1949) by Carlos A. Bertameu, or *Francisco P. Moreno: Arquetipo de argentinidad* (*Francisco P. Moreno: Archetype of Argentineness*, 1954) by A. D. Ygobone.[38] Of these biographies, the one that stands out the most is José Ingenieros's exhaustive study, *Las doctrinas de Ameghino* (*The Doctrines of Ameghino*, 1919).

It is interesting to note that the animosity between Ameghino and Moreno—which led the former to forbid the latter from entering the Museum of La Plata—was rekindled in later years by the writers of the scientists' respective biographies. Early "Ameghinism" was based on the accolades of Ingenieros, Lugones, and Rojas, and was marked by the controversy over the scientist's Italian nationality, as well as the clash between its adherents' socialism on the one hand and Catholic groups on the other. Thus, beginning in 1915 and especially following the first publications of the Socialist Party, Ameghino would become the symbol of progressive and secular science. His defense of Darwinism and his rebelliousness toward authority (particularly that of Burmeister) only bolstered his reputation as a revolutionary scientist.

Upon entering the sciences, Ameghino was subject to the field's rather strict rules, whereby the rise and fall of its members was determined by political favoritism. In the specific case of Ameghino, it is important to keep in mind that at this time Genoese immigrants in Argentina were a growing threat to the position of the local oligarchy.[39] Unlike Moreno and Holmberg, the brothers Carlos and Florentino were not the children or in-laws of the "old criollo" families, nor did they attend the same schools or move in the same social circles as them. In a suggestively titled article, "De la santidad laica del científico Florentino Ameghino y el espectáculo de la ciencia en la Argentina moderna" ("On the secular sainthood of the scientist Florentino Ameghino and the spectacle of science in modern-day Argentina"), Irina Podgorni recalls the Catholic "Patriotic League." This was a group which was opposed to the University Reform and saw in Moreno an alternative figure to that of Ameghino. It was on account of this that, for the Socialists, Moreno came to be known as an anti-evolutionist associated with reactionary ideals.[40] Since the dawn of the Darwinist controversy in Argentina, Catholicism had come out against the theory of evolution, which did not recognize the Divine Plan in the creation of humanity. In 1880 José Manuel Estrada (1842–1894), in his speech, "Naturalism and Education"—an excerpt of which is included in this anthology—came out against Darwin, his "villainous consequences," and the "laughable pride" of his followers.

The heated controversy between supporters of Ameghino and Moreno rekindled the scientific ferment depicted by Holmberg in 1874 in *Two Parties in Conflict*. The novel, part of which appears in this anthology, is a fictional, satirical account of a visit to Buenos Aires by Darwin at the invitation of the Argentine Scientific Society. Through a surgical operation performed on an Akka man (African Pygmy) in the Teatro Colón, Darwin discovers the missing link between man and monkey. The city of Buenos Aires euphorically hails Darwin's arrival, and welcomes the corroboration of his theory as a "philosophical revolution of incalculable transcendence both for science as well as the nation's politics."[41]

In this way, to be a follower of Darwin became associated with the ideology of progress, secularism, and in some cases with socialism in liberal Argentina.[42] This is true for Ameghino, Rojas, Ingenieros, and even Lugones, an excerpt from whose book on Ameghino appears in this anthology, along with his fantasy story "Yzur." Much like Holmberg, the author of *Fuerzas extrañas* resorts to fantasy to explore the period's common and contradictory mixture of spiritualism, mesmerism, and science.[43]

I would like to underscore that the fundamental difference between the two main authors in this part—Ameghino and Moreno—on the one hand, and those included in the third part of this anthology on the other, is the way they were appraised after their time on Argentina's intellectual scene was over. What writers emphasized about both Ameghino and Moreno was that their work lacked pragmatism. These two naturalists' studies in paleontology did not revolve around the economic interests that, by the end of the nineteenth century, had come to guide the practical application of the sciences in Argentina. The historian Juan Carlos Zuretti points out that the focus of most scientific research at this time was the burgeoning industrial sector, which was being nourished by the 20,000 leagues (approximately 60,000 miles) of land conquered from the Indians, by the growing immigrant labor force, and by the proliferation of farming colonies.[44] At the same time public works like railroads, public health projects, and the development of the ports promoted a growing interest in technological studies at the expense of "pure" science, particularly between 1890 and 1916. In the words of Zuretti, "people praised material progress, loved the accumulation of wealth, and had little love for the pure values of science."[45]

José Babini referred to this moment as "the crisis of scientific progress of the 90s." With this crisis scientific research marched "to the beat of the deluge of immigration" and followed "behind utilitarian and material desire."[46] It is no coincidence then that Moreno and Ameghino's painstaking dedication to a non-industrial or "pure" science was seen as a model of patriotism, since the type of science that these two naturalists sought to institutionalize was based not on wealth, but on the creation of a national heritage.

Pierre Bourdieu uses the term "nomothète" to refer to those people who founded the modern field of the arts, those who, like Baudelaire, galvanized the task of the poet or artist through their adherence to the ideology of "art for art's sake." One is a nomothète insofar as he or she believes that the reward for his or her creativity is artistic creation itself, removed from marketing strategies and the laws of the bourgeois or "mass" market. The nomothète, for Bourdieu, attempts to "impose within the field as the legitimate view on the field, the fundamental law of the field, the principle of vision and division (*nomos*) defining the artistic field *as such*, meaning as the site of art as art."[47] To secure this authority, the nomothète must prove him or herself through suffering, poverty, and exclusion from fame. Extrapolating this model from the arts to the sciences, travelers and naturalists like Ameghino and Moreno could be called scientific nomothètes, at least in the way that their biographers present them. In the case of Moreno, his career is portrayed as a sort of Stations of the Cross, sacrificing himself in inhospitable frontier lands with Indians lying in wait. The followers of Ameghino, for their part, emphasize that he worked in poverty with no institutional reward.

Through these ideological apparatuses concerning the creation of heritage and the nation's scientist-heroes, the Darwinism to which they subscribed lost in part its characteristic tendency to legitimate progress as a "struggle for life" and "survival of the fittest" in the campaign to modernize the nation.

PART III: DARWIN AND THE SHAPING OF ARGENTINE IDENTITY: HISTORICAL-SOCIOLOGICAL STUDIES (BUNGE, RAMOS MEJÍA, INGENIEROS)

Although the decades from 1880 to 1910 saw the superposition of different ideologies and philosophical doctrines, among them vitalism, decadentism, and the modernist spirituality of Rodó's *Ariel*, the period's dominant trend was without a doubt positivism.[48] One could even go so far as to say that the aforementioned philosophical trends arose as a reaction and response to positivism.

It has been pointed out that positivism was the ideology that accompanied and legitimated the economic modernization of Argentina and its state policies, aiding the country's entrance into the world market. The centralization of the state, coupled with the rise of positivism, homogenized (bureaucratically and discursively) the diverse social structures (rural migrant workers in the city, immigrants, politicized labor, delinquents, women, and homosexuals), measuring them up and comparing them in order to make them governable.

According to José Luis Romero, during Roca's first presidency (1880–1886) some 483,000 immigrants entered Argentina. Population growth was accelerated, as an average of more than 80,000 immigrants poured into the country per year, reaching a figure of 261,000 immigrants in 1889 alone, and still more in 1906. In the year 1930 the population increased by 223,000 inhabitants. This growth stimulated economic development, stockbreeding, and agriculture, and was a boon to industry as well: by 1895 there were already 24,114 industrial establishments in the country, which employed a total of 175,000 workers.[49]

During this time Buenos Aires underwent abrupt economic and demographic changes. There was an outward image of prosperity, thanks to a number of capital investments, along with British financing of public services such as railroads, streetcars, telephones, natural gas, sewer systems, and government loans. However, this apparent prosperity masked "flagrant economic and social contrasts."[50] On the north side of the city of Buenos Aires sumptuous mansions were built, while in the south people piled into unsanitary tenements.

Immigrants in Argentina were generally illiterate Italians (Piedmontese, Genoese, Neapolitan, Calabrian) and Spaniards (Galicians, Basques, Andalusians), though they hailed from a far-flung variety of other origins as well: French, Jews from Russia and Poland, Arabs and Syrio-Lebanese, Turks, etc. These immigrant populations fanned out geographically, extending the city limits of Buenos Aires and creating a new poor quarter. To this we must add the slow and steady migration from the interior to the capital, caused by the reorganization of rural property with the introduction of the wire fence.[51]

The unavoidable presence of these masses of migrants caught the attention of the "old criollo" elite. Proof of their economic successes in some sectors sparked both criticism and a need to explain their social and moral structures, given that these people whom they pejoratively called "gringos"[52] were beginning to displace the criollo class in economic production through their determination and perseverance.[53] The elites attempted to assimilate immigrants into the mainstream progress of the nation, while at the same time holding on to their own hegemonic status.

This desire, at once liberal and conservative, brought with it an obligation to explain the "local evils" posed not only by the immigrant bourgeoisie, but also by labor organizations and the appearance of the Radical Civic Union (*Unión Cívica Radical*) on the political landscape. The goal was to normalize the connections between the state apparatus and society. Positivism was the tool that was used to deal with the new multitudes, and was likewise employed to understand not just the nation's present, but its past as well.[54]

Oscar Terán has outlined the links between Herbert Spencer's (1820–1903) evolutionist philosophy and the positivist method of Auguste Comte (1798–1857), and the way in which both were used to explain national problems.[55] When making this connection it is important to emphasize that positivism dialogued not just with biological Darwinism, but with social Darwinism as well. The latter was understood to be an interpretation of evolutionism that supported "the survival of the fittest," meaning the criollo elite, in the socioeconomic sphere. Even though—as Soler notes—both Bunge and Ingenieros exhibited a deliberate "return to biological Darwinism," this did not cease to be a clear "reinterpretation of Darwin."[56]

The key to distinguishing between social and biological Darwinism in this period is the association between evolution and progress. For Darwin, evolution is probabilistic, in that species' mutations are not aimed at a predetermined end, and changes in varieties are produced arbitrarily. This makes it possible for the better adapted to have higher chances of survival in an environment which likewise mutates without any teleological aim. For Spencer, on the other hand, evolution is deterministic or mechanistic, given that each stage of human society is the logical consequence of a prior stage, to which it is in turn "superior."[57]

In his study on positivism in Argentina, Carlos Mayo calls Argentina the "Spencerian province."[58] For Mayo it is not difficult to understand how social Darwinism spread throughout Argentina from the time of Roca until the centennial, given that from the time of the Generation of 1837's Romantics, the idea of civilization's progress had been an obsession for the intellectual elite. Sarmiento himself recognizes this in his speech on Darwin: "Spencer and I get along well, because we walk the same path."[59]

Mechanist Darwinism's absolute faith in the progress of societies, along with the materialist vision of science in the face of the obscurantism of religion, led Spencer to fuse together the debates that were being held not just among sociologists, doctors, and intellectuals, but among the politicians in Parliament as well.

Progress was seen as an inevitable law of history across the board in the Miguel Juárez Celman administration's liberal politics, and this belief brought with it the idea that the state should not interfere with personal capital, and thus allow for the struggle for life to bring about the "survival of the fittest."This drove the state to sell off public lands and make concessions to the railroad companies, which in the end led the country into the catastrophe of 1890, when the feverish activity of the stock market brought state and personal finances to the "edge of the abyss."[60]

The texts by doctors, lawyers, and politicians selected for this anthology —José Ingenieros, José María Ramos Mejía, Carlos O. Bunge, and Sarmiento—propose an evolutionist reading of the past and present of Argentina's culture, through an analysis of the psychology and sociology of the "mestizo" or "half-breed," the gaucho living under the caudillos,[61] and the immigrant during the period of mass immigration. These authors reinterpret Darwin's original proposal, though they also, as has been said, feed off of Spencer, Le Bon, and others' mechanistic and social theories of Darwinism.

In relation to Ramos Mejía and Ingenieros, critics have studied their use of medical discourse as a tool for social analysis,[62] and have underscored their use of metaphors like the "sick body" and the "mental and moral pathologies" to characterize social prototypes. This anthology collects a group of texts in which the medical discourse and its metaphors are teamed up with the theory of evolution in order to explain how the masses are inferior to the ruling elite, which is made up of politicians and landowners, as well as intellectuals.

José María Ramos Mejía (1849–1914) carried out a vast body of work as a doctor, politician, and essayist. He was the creator of the Public Healthcare Program, the Department of Sanitation, and the university's Department of Neuropathology, and was a member of the Medical Circle. This was the same Medical Circle that organized the homage to Darwin in 1882 following his death, at which Sarmiento and Holmberg gave the two speeches that are included in Part II.

In *The Argentine Masses* (1899), an excerpt of which is included here, Ramos Mejía takes on the problem of the masses' governability throughout the country's history. For the author, in the gaucho and immigrant multitudes the ability to reason is lost and the individual is "primitivized," returning to prior stages in the evolution of the species. If the masses lose their rational faculties, the issue is how to turn the masses into citizens, how to bridge the gap between the multitudes' production of wealth for the nation and their ability to be full citizens. This debate is connected with the controversy over universal suffrage and the ruling elite's opposition to it. Gino Germani points out that in 1895 more than 50 percent of male adults were excluded from political society. This exclusion was based on the belief that the masses "were incapable of rationally differentiating between the different options on the ballot."[63]

Ramos Mejía goes on to sketch out a descriptive typology of the masses, which he divides up into the *oaf* (*guarango*), the *riffraff* (*canalla*), the *hick* (*huaso*), and the *golden bourgeois* (*burgués aureus*). According to Ramos Mejía, "[i]n social paleontology, the *oaf* represents one of those vertebrates that will be searched out with curiosity by the sociologists of the future, in order to establish the succession of biological forms in our evolution." At the same time the prosperous immigrant, or the *golden bourgeois* "in large numbers, will be formidable if our national education system does not modify him with the scrub brush of culture and the percolation of other ideals that might hold him back from his hasty ascension to the Capitol." Ramos Mejía then proposes that public education is the right way to treat and cure the sick social psyche of the immigrant and adapt him to the civic environment.

In *Our America* (1903) Carlos O. Bunge (1875–1918) takes this biological positivism to the extreme, initiating a sort of pre-Galdonian eugenics. In the essay included in this anthology, Darwinism is employed to explain the processes of atavism, according to which "*a cross between two distinct varieties of a species reproduces the specific ancestral form.*" According to Bunge, hybrids and crossbreeds degenerate back to the primitive characteristics of prior species in their lineage, though in the end he does not reach a conclusion as to whether hybridism or cross-breeding between the human races is altogether harmful. He does, however, venture that if the idea of biological atavism can indeed be applied to the human races, then "the crossbreed will tend to reproduce a primitive form of man, or at least one that is ancient and pre-Christian." Likewise, Bunge uses the principle of "correlation," a phenomenon studied in the animal world by Darwin himself, to conclude that there is a "correlation" between "crossbreeds'" physical and psychological characteristics, namely "*psychological discord, degenerative semi-sterility, and lack of moral sense.*" Further on he exhibits yet another brand of racism in his long-winded description of the mulatta woman, who for him is the ideal object of desire for the white man.

Bunge's "Darwinist" racism is evident in every paragraph, and he goes so far as to encourage the use of anthropological methods to watch for and detect how assimilated "crossbreeds" cover up their hidden traits. His evolutionist doctrine leads him to bless alcoholism, smallpox, and tuberculosis for having decimated the indigenous and black populations of Buenos Aires, thereby "Europeanizing it." Bunge contrasts this, negatively, with the interior of the continent, where "the tropical temperatures, the difficulties of communication with the metropolis, and the culture and beauty of the indigenous races spread and increased crossbreeding up to the period of independence."[64]

For his part, José Ingenieros (1877–1925) combined aspects of these positivist and Darwinist ideas with socialist doctrines and laws of economics. He carried out his research in psychiatry and criminology as the clinical director at the Buenos Aires Police Department's Observation of Outcasts Service, and at the national penitentiary's Institute of Criminology. Ingenieros understood sociology as a *natural science*, and set out to test a genetic interpretation of Argentina's evolution, based on the premise that all sociological phenomena are merely the combination of a series of physical, biological, and psychological circumstances. Man, for Ingenieros, "is not an aerolite fallen to earth through some whim of the supernatural forces," but rather "a living creature, nothing more. . . . He fights for his life in order to satisfy basic and indispensable needs: the preservation of the individual and the reproduction of the species." As opposed to Ramos Mejía and Bunge, Ingenieros thought that humanity, as a biological group, did not have a mission in the universe and lacked absolute moral principles. For Ingenieros, "[t]he spring that sets man's social activity—his conduct—in motion is the sum of his needs; his consciousness of the latter—subjected to rigorous determinism—is merely the apparent motive behind all individual or collective actions." Thus, Ingenieros believed that biological sociology should replace Spencerian organicism.

Ingenieros employed this biological perspective on social evolution to depict imperialism as the leadership of the fittest in an international struggle for survival. Ingenieros saw expansionism as a phenomenon which was subject to natural and scientific laws outside the realm of moral judgment, and, because of his confidence in the elite leadership of the "fittest," he predicted that Argentina would become the supreme political power in South America, one which would be able to oppose North America's United States on equal footing.

Sarmiento's *Conflict and Harmonies of the New World Races* (1889) evidences a monumental bibliographic research employed to scientifically explain the history of the New World population in terms of race, biology, and psychology, and also represents the author's drive to analyze the similarities and differences between North and South American histories. Sarmiento's admiration of the United States of America is well documented.[65] In this sense, Sarmiento and Ingenieros can be read together in their hopes/plans for an imperialist future for Argentina based on their evolutionist interpretation of history, and the northern country's model of development.

For Sarmiento and Ingenieros, positivism and Darwinism along with the United States' modernization were synonymous with progress, and were models to follow in the planning of a nation, racially and historically. From different perspectives and disciplines, all authors in this anthology envisioned a promising future for Argentina. However, overwhelming challenges on the global stage were to come in the following decades.

NOTES

1. Charles Darwin, *From so Simple a Beginning: The Four Great Books of Charles Darwin*, ed. Edward Wilson (New York: Norton, 2006), 151.

2. Alan Moorehead, *Darwin y el Beagle (1831–1836)* (Barcelona: Ediciones del Aguazul, 2002), 119.

3. J. Vernon Jensen, *Thomas Henry Huxley: Communicating for Science* (Newark: University of Delaware 1991), Chapter 3.

4. Sandra Gasparini, "Introducción," in *Dos partidos en lucha*, by Eduardo Holmberg (Buenos Aires: Corregidor, 2005), 35.

5. See Adolfo Prieto, *Los viajeros ingleses y la emergencia de la literatura argentina 1820–1850* (Buenos Aires: Sudamericana, 1996); Jean Franco, "Un viaje poco romántico: Viajeros británicos hacia Sudamérica," *Escritura* 7 (1979): 129–42.

6. Roberto González-Echevarría, *Myth and Archive* (New York: Cambridge University Press, 1990), 93–97.

7. Ibid., 97.

8. José Babini, "Breve historia de la ciencia argentina," in *La ciencia en la Argentina. Perspectivas históricas*, ed. Miguel Asúa (Buenos Aires: CEAL, 1993), 34.

9. Aníbal Ford, "Darwin, Fitz-Roy y los intereses ingleses en el Atlántico Sur: Ensayo de contrainformación cultural," in *Desde la orilla de la ciencia: Ensayos sobre identidad, cultura y territorio*, ed. Aníbal Ford (Buenos Aires: Punto Sur, 1988), 72.

10. Ibid., 73.

11. See Mary Louise Pratt, *Imperial Eyes* (London: Routledge, 1992); David Spurr, *The Rhetoric of Empire* (Durham: Duke University Press, 1993); Ricardo Salvatore, ed., *Culturas imperiales* (Rosario: Beatriz Viterbo, 2005); Edward Said, *Orientalism* (New York: Pantheon, 1978).

12. Bruno Latour, *Science in Action: How to Follow Scientists and Engineers through Society*. (Milton Keynes: Open University Press, 1987).

13. For a pertinent discussion on the topic of social construction of science and Latour's constructivist model, see Keith M. Parsons, *The Science Wars: Debating Scientific Knowledge and Technology* (Amherst: Prometheus, 2003); Frédéric Vandenberghe, *Complexités du posthumanisme: Trois essais dialectiques sur la sociologie de Bruno Latour* (Paris: L'Harmattan, 2006); Bill Brown, *Things* (Chicago: University of Chicago Press, 2004); Craig Hanks, *Technology and Values: Essential Readings* (Chichester, U.K. and Malden, MA: Wiley-Blackwell, 2010); and David Cayley, *Ideas on the Nature of Science* (Fredericton: Goose Lane Editions, 2009).

14. Marcelo Montserrat, "La sensibilidad evolucionista en la Argentina decimonónica," in *La ciencia en Argentina entre siglos*, ed. Marcelo Monserrat et al. (Buenos Aires: Manantial, 2000), 206.

15. Darwin's presence is a constant in Hudson's work. In his famous "Biography of the Vizcacha," in *A Naturalist in La Plata* (1892), Hudson cites Darwin to support what is most uncommon and interesting about the rodent: its ancientness. The vizcacha, according to Darwin in *On the Origin of Species*, preserves traits inherited from its ancient ancestor: a marsupial. Thus the animal could be considered a "missing link" species between marsupials and rodents, the signs of which disappeared over the course of millennia. The vizcacha has, then, an incomparable scientific prominence, as it is evidence of the common origin of marsupials and rodents, and yet another proof of the theory of evolution.

16. Julio Orioni and Fernando Rocchi, "El darwinismo en la Argentina," *Todo es historia* 228 (1986): 10.

17. Domingo F. Sarmiento, *Vida y escritos del coronel D. Francisco J. Muñiz* (Buenos Aires: Félix Lajouane, 1885), 10.

18. Walter Mignolo, *The Dark Side of the Renaissance* (Ann Arbor: University of Michigan Press, 1995); Anthony Pagden, *European Encounters with the New World: From Renaissance to Romanticism* (New Haven and London: Yale University Press, 1993).

19. TN: In Argentina and other parts of Latin America "criollo," sometimes translated as "creole" in English, refers to native-born descendents of European and particularly Spanish settlers. I have chosen not to translate it as "creole" because the standard meaning of that term in English is quite different, i.e., a person of combined European and African descent.

20. Rosa Fung de Lanning, "Las ideas evolucionistas en las interpretaciones arqueológicas," *Revista del Museo Nacional* 32 (1963): 203.

21. Nigel Leask, *Curiosity and Aesthetics of Travel Writing* (New York: Oxford University Press, 2002).

22. Adriana Novoa, "The Dilemmas of Male Consumption in Nineteenth-Century Argentina: Fashion, Consumerism, and Darwinism in Domingo Sarmiento and Juan B. Alberdi," *Journal of Latin American Studies* 39 (2007): 771–96; Evelyn Fishburn, *The Portrayal of Immigration in Nineteenth Century Argentine Fiction (1845–1902)* (Berlin: Colloquium Verlag, 1981). It is interesting to note that both Sarmiento's *Conflict and Harmonies of the New World Races* and Alberdi's *Luz del*

Día's Pilgrimage were republished in José Ingenieros's collection *La cultura argentina* in 1916, which indicates an ongoing debate in the intellectual field. Ingenieros considered Sarmiento's a foundational text for Argentinean scientific sociology, and decided to republish *Conflict and Harmonies* before the canonical *Facundo: Civilization and Barbarism* (1845). See Fernando Degiovanni, *Los textos de la Patria* (Rosario: Beatriz Viterbo, 2007), 277.

23. Here Sarmiento presents the first phase of Moreno's museum, which will later be moved to La Plata, and was at that time located in the Teatro Colón. Domingo F. Sarmiento, "Mundos prehistóricos: Viaje aéreo a través del Museo Prehistórico de Moreno," in *Obras* (Buenos Aires: Imprenta Mariano Moreno, 1899), 135–45.

24. Beatriz González-Stephan, "Showcases of Consumption: Historical Panoramas and Universal Expositions," in *Beyond Imagined Communities*, ed. Sara Castro and John Chasteen (Baltimore: Johns Hopkins University Press, 2003), 225–38. See also Beatriz González-Stephan and Jens Andermann, *Galerías del progreso: Museos, exhibiciones y culturas visuales en América Latina* (Rosario: Beatriz Véterbo, 2006).

25. Álvaro Fernández Bravo, "Latinoamericanismo y representación: Iconografías de la nacionalidad en las exposiciones universales (París 1889–1900)," in *La ciencia en la Argentina entre siglos*, ed. Marcelo Montserrat (Buenos Aires: Manantial, 2000), 177.

26. Babini, "Breve historia de la ciencia argentina," 37–40.

27. Marcelo Montserrat, "La mentalidad evolucionista: una ideología del progreso." In *Ciencia, Historia y Sociedad en la Argentina del siglo XIX*, ed. Marcelo Montserrat, (Buenos Aires: CEAL, 1993), 45.

28. Darwin had only been elected corresponding member of the botanical section of the *Académie des Sciences de Paris*, and not without internal conflict at the Academy. On the other hand, the National Academy of Córdoba awarded him the title of honorary member two days before the French Academy did (on August 3rd, 1878). See Marcelo Montserrat, "La mentalidad evolucionista," 45.

29. It is important to remember that here "Creation" does not admit any supernatural intervention or divine design. Creation means "existence": all that exists in the universe. Many pre-evolutionist and even evolutionist authors used the term "Creation" without any supernatural meaning, i.e., *Vestiges of the Natural History of Creation* (Robert Chambers, 1844), *The History of Creation* (Ernst Haeckel, 1868). Burmeister denied any ex nihilo creation and accepted spontaneous generation just as a hypothesis. Leonardo Salgado and Pedro Navarro Floria, "Hermann Burmeister y su *Historia de la creación*: Idealismo, materialismo y empirismo en el credo de la primera ciencia argentina," in *Patagonia: Ciencia y conquista; La mirada de la primera comunidad científica argentina*, ed. Pedro Navarro Floria (Neuquén: CEP, Universidad Nacional del Comahue, 2004), 37–60. For Burmeister, the history of creation followed a series of geological revolutions characterized by extinction and volcanic movement that ended up improving the conditions for human existence.

Burmeister was antireligious and specifically, he opposed Catholic miraculous explanations of natural phenomena. Irina Podgorni reminds us that on one occasion, Burmeister attacked Ameghino's stubbornness by comparing it with the dogmatism of some religious beliefs. See Irina Podgorni, "Prólogo," in *Patagonia: Ciencia y conquista; La mirada de la primera comunidad científica argentina*, ed. Pedro Navarro Floria (Neuquén: CEP, Universidad Nacional del Comahue, 2004), 24.

30. Quoted in Montserrat, "La mentalidad evolucionista," 35–36.

31. Salgado and Floria, "Hermann Burmeister y su *Historia de la creación*."

32. See Rosaura Ruiz Gutiérrez and Francisco Ayala, *De Darwin al DNA* (México: Fondo de Cultura Económica, 2002).

33. Cristina Mantegari, *Germán Burmeister: La institucionalización científica en la Argentina del siglo XIX* (Buenos Aires: Jorge Baudino Ediciones/UNSAM, 2003), 178.

34. James Clifford, *The Predicament of Culture* (Boston: Harvard University Press, 2002), 223.

35. Ernesto Livón-Grosman, *Geografías imaginarias: El relato de viaje y la construcción del espacio patagónico* (Rosario: Beatriz Viterbo, 2003), 103–51.

One example of this is the letters Moreno wrote to different figures asking them to free the chiefs Inacayal and Joyel, friends of his from his travels to Patagonia, who had been taken prisoner following the occupation of their territories. See Francisco Moreno, *Reminiscencias del Perito Moreno* (Buenos Aires: El Elefante Blanco, 1997), 261–63.

36. Ricardo Rojas, *Los arquetipos* (Buenos Aires: Roldán, 1922), 206–7.

37. Although in the case of Ameghino one is obliged to mention his brother Carlos (1865–1936), who collaborated tirelessly with Florentino, and was a true leader on many heroic expeditions through inhospitable territories. See Florentino Ameghino and Carlos Ameghino, *Reseñas de la Patagonia: Andanzas, penurias y descubrimientos de dos pioneros de la ciencia* (Buenos Aires: Ediciones Continente, 2006).

38. *Holmberg: El último enciclopedista* (*Holmberg: The Last Encyclopedist*, 1952) by Luis Holmberg could also be considered in this same vein.

39. Irina Podgorni, "De la santidad laica del científico Florentino Ameghino y el espectáculo de la ciencia en la Argentina Moderna," *Entrepasados: Revista de Historia* 6, no. 13 (1997): 38.

40. Ibid., 54.

41. It may be useful to clarify that in the novel the name "Rabianist" is not synonymous with "creationist," but rather a group that defended conservative social values, which were not necessarily connected to the biblical story of creation. Catholicism as a religious doctrine is not put in question in the novel.

42. Other members of the Buenos Aires elite, like Miguel Cané in his review of *Two Parties in Conflict*, simply refused to accept the ideas of the evolution of species and the relationship between man and ape out of a simple question of "good taste." See Montserrat, "La mentalidad evolucionista," 44.

43. Marta Fabbri, "Inserciones de lo fantástico en las *Fuerzas extrañas*," in *El relato fantástico rioplatense en el siglo XIX*, ed. Enriqueta Morillas Ventura (Neuquén: Universidad Nacional del Comahue, 2005), 53–73.

44. Juan Carlos Zuretti, *Historia de la cultura argentina: Arte, ciencia* (Buenos Aires: Itinerarium, 1964), 259.

45. Ibid., 271.

46. Quoted in Gregorio Weimberg, *La ciencia y la idea del progreso en América Latina, 1860–1930* (México: Fondo de Cultura Económica, 1996), 77.

47. Pierre Bourdieu, *The Rules of Art* (Stanford: Stanford University Press, 1996), 223.

48. Oscar Terán, *Positivismo y nación en la Argentina* (Buenos Aires: Puntosur, 1987), 11.

49. José Luis Romero, *Las ideas políticas en Argentina* (Buenos Aires: FCE, 1959), 175.

50. Eduardo Romano, "Fray Mocho: El costumbrismo hacia 1900," *Historia de la literatura argentina*, vol. 2 (Buenos Aires: CEAL, 1980), 265.

51. Ibid., 265–6.

52. TN: Applied in Argentina to foreigners generally.

53. Romero, *Las ideas políticas*, 181.

54. Oscar Terán reminds us that positivism circulated in the climate of general "Argentinocentric" optimism, which displayed great confidence in national progress, unprecedented economic expansion, growing social mobility, and cultural modernization stimulated by the state. See Oscar Terán, *Positivismo y nación en la Argentina* (Buenos Aires: Puntosur, 1987).

55. Carlos Mayo says:

> Without a doubt, and their differences aside, there are certain common assumptions in the thinking of [Spencer and Comte]: 1) the scientific method and the sciences in general offer the best and most certain instrument of knowledge; 2) positivism, as a theory of knowledge, does not acknowledge any other reality aside from actions and their consequences; for this reason it rejects all aprioristic and metaphysical knowledge; 3) society mirrors nature, and can share with it the same type of predictable and inevitable laws, or itself be natural, organic, and biologically determined; 4) history tends, through a series of successive stages or through an evolutionary process, towards progress.

Carlos Mayo, *El positivismo en la política argentina 1880/1906* (Buenos Aires: CEAL, 1988), 2.

56. Soler writes:

> Bunge sought to apply this [biological] determinism in the domain of sociology using a psycho-social conceptualization, which was based on the evolution of the most basic biological principles (the protozoan's adaptation through the principle of pleasure and pain). Ingenieros's connection between biology and society is based on an economic determinism that he sees as dependent upon the human species's sociogenetic evolution.

Ricaurte Soler, *El positivismo argentino* (Panamá: Imprenta Nacional, 1959), 195.

57. The idea that societies advance progressively was fairly widespread, and was promoted not just by Spencer but by others as well, among them Thomas Malthus and Thomas Huxley. For Malthus, overpopulation would bring about food scarcity and with it, the elimination of the weak. For his part, Huxley, who was a stalwart defender of Darwin in the Oxford Debate and throughout his life, supported the idea that the Nordic races' ability to adapt to a harsh climate had made them superior to the rest, since it had swiftly wiped out the weak among them.

58. Mayo, *El positivismo en la política,* 4.

59. Domingo F. Sarmiento, *Obras*, vol. 22 (Buenos Aires: Imprenta Mariano Moreno, 1899).

60. Romero, *Las ideas políticas*, 200–201.

61. TN: The Latin American "caudillo" is a rural political leader, a sort of political boss who exerts a power outside of and sometimes in opposition to the power of the state. In the nineteenth century many gauchos lived in areas controlled by these figures.

62. Gabriela Nouzeilles, *Ficciones somáticas* (Rosario: Beatriz Viterbo, 2000); Jorge Salessi, *Médicos, maleantes y maricas* (Rosario: Beatriz Viterbo, 1995); Rafael Huertas, *El delincuente y su patología: Medicina, crimen y sociedad en el positivismo Argentino* (Madrid: Consejo Superior de Investigaciones Científicas, 1991).

63. Terán, *Positivismo y nación*, 22.

64. For a discussion on eugenics, see Joseph Graves, *The Emperor's New Clothes: Biological Theories of Race at the Millennium* (New Brunswick: Rutgers University Press, 2001); Elof Axel Carlson, *The Unfit: A History of a Bad Idea* (Cold Spring Harbor: Cold Spring Harbor Laboratory Press, 2001); Philip Kitcher, *In Mendel's Mirror: Philosophical Reflections on Biology* (Oxford and New York: Oxford University Press, 2003); and John Jackson and Nadine Weidman, *Race, Racism, and Science: Social Impact and Interaction* (Santa Barbara: ABC-CLIO, 2004), among others.

65. Domingo F. Sarmiento, *Viajes por Europa, África y América* (Madrid: Fondo de Cultura Económica, 1996); Halperín Donghi et al., *Sarmiento: Author of a Nation* (Berkeley: University of California Press, 1994).

I

THE FIRST READERS OF *ON THE ORIGIN OF SPECIES* IN ARGENTINA

The Dialogue between Darwin and Muñiz, and the Controversy with Hudson

At twenty-four, from 1869 to 1870, Hudson began a scientific exchange with the Smithsonian Institute in Washington, specifically with the ornithologists Spencer Fullerton Baird and Phillipe Lutley Scatler. This exchange was facilitated by one of the main anti-Darwinists in Argentina, Hermann Burmeister, then the director of the Museum of Natural Sciences of Buenos Aires. Hudson's exchange at that time consisted of embalmed specimens of native birds that he sent to London, along with a detailed description of the species and their behavior. As a result, the metropolis named at least two of the bird species that he sent in his honor: *Knipolegus hudsoni* (Hudson's black-tyrant) and *Asthenes hudsoni* (Hudson's canastero). Hudson's writings were edited by Scatler and published in the *Proceedings of the Zoological Society of London*. In this way, metropolitan science had access to Hudson's "firsthand" knowledge, even before the local knowledge networks had the resources necessary to disseminate this information within Argentina's scientific circles. A few years later, in 1873, Hudson moved to London permanently.

Hudson was born into a family from the United States that had moved to the pampas, where he was raised among Irish tutors and English gauchos. This background gave Hudson early access to *On the Origin of Species*: On a trip to England, Hudson's older brother obtained a copy for him, which he read avidly as an adolescent.

"The Pampas Woodpecker," is one of the most important of those works compiled by Scatler in the aforementioned London periodical, and is proof of this sustained debate. Although Hudson makes some strong allegations against Darwin, the controversy between the metropolitan and the local never transcended the epistolary debate. If read carefully, Hudson's observations do not, in fact, refute Darwin's theory, but rather constitute his own minor contribution to it.

Based on his observations of the pampas woodpecker, Darwin had concluded that it constituted the strangest case of this family he had found during his travels aboard the *Beagle*, given that the bird's traits, despite the fact that it lived in areas devoid of trees, had not been modified. Darwin found this odd because, in this new setting, nature had apparently left the bird intact, a fact which would run counter to the ideas of the variability of the species and the survival of the fittest. Hudson counters that the pampas do indeed boast the *ombú* tree, and in areas not far from where Darwin made his observations. This, he asserts, is the explanation behind the unmodified woodpecker.

For Hudson, Darwin's conclusions rest upon observations made over the course of just a few days' travel through the area. Because Hudson lived in the pampas, and his observations were regular and constant, he positions himself as the best authority to describe the woodpecker. Here, local experience becomes symbolic capital. Despite Darwin's well-known reluctance to engage in public controversy, on this occasion, he decides to take part in the debate and rebut Hudson's accusations on both scientific and moral terms. Darwin not only upholds his own observations, but also quotes from another local authority in the matter, the ornithologist Félix de Azara (1746–1820), who lived in the area for almost three decades, and was author of the most important book on birds at the moment, *Apuntamientos para la historia natural de los páxaros del Paraguay y del Río de la Plata* (*Notes for a Natural History of the Birds of Paraguay and Río de la Plata*, 1802).

Interesting is to note that in one of his later works, *Birds of La Plata* (1920), already living in England, and quoting Darwin and Azara, Hudson expressed his mixed feelings about his effrontery, or disobedience to the masters. Nevertheless, the naturalist mentioned once again the woodpecker that lived in the *ombú* tree, and that he met in the pampas.

MUÑIZ AND DARWIN: THE *ÑATA* COW

Muñiz's contributions earned him recognition as a paleontologist ahead of his time in a pre-scientific Argentina. Sarmiento points out that "with Muñiz there began, in this country, a scientific and literary movement aimed at studying ourselves and the country in which we live" (*Vida y escritos*, 10). Muñiz was a military doctor in Cepeda and in the Paraguay war, was a university professor and chair of the department of "theory and practice of birth, childhood diseases, and medical jurisprudence," and worked to bring about health reforms in Argentina. Muñiz carried out excavations throughout his fifteen-year residency in Luján, and his numerous finds enriched collections not just in Buenos Aires, but in Spain, Sweden, and France as well, as a significant portion of the extremely valuable fossils that Muñiz offered to President Rosas, in hopes of founding a natural history museum, were to end up in foreign hands.

When Darwin traveled through Argentina he did not have the opportunity to meet Luján's doctor and collector of fossils. However, the observations on the *ñata* cow that Muñiz provided through Darwin's friend Enrique Lumb proved crucial to the theory of evolution and to Darwin's studies into the mutations of domesticated animals. Included in this section is the letter that Darwin sent Muñiz thanking him for his descriptions of the *ñata* cow and the Muñi-Felis fossil.

Muñiz's study of the Muñi-Felis fossil triggered a debate over the size of American fauna. European naturalists such as Buffon and DePaw had claimed that both the fauna and the people of the New World were inferior in size to those of the Old World, seeing a clear parallel between the size of the "primitive" regions' species on the one hand, and their supposedly childlike character on the other, as opposed to the "civilized" ones, that is to say, those that were Indo-European, Arabian, or Chinese in origin. Thus, in the same way that the New World puma was smaller than the Old World lion, the Amerindian man and woman were wrongly perceived to be homologously childlike, due to their supposed lack of any non-iconic writing systems or sophisticated technology, such as alphabetical writing.

What the Muñi-Felis demonstrated—according to Muñiz—was that there had once been a big cat whose size was greater than that of Old World felines, meaning that the fossil contradicted prejudices and beliefs about New World wildlife. In doing so, Muñiz exhibits a criollo's drive to redeem the Americas' natural world before the emerging field of paleontology, and that is one of the main reasons why Sarmiento becomes interested in his studies and welcomes him into the national pantheon of scientists.

SARMIENTO ON MUÑIZ AND DARWINISM

The first to describe the *ñata* cow was Muñiz, who then sent his descriptions to Darwin. The *ñata* cow, Sarmiento points out in the article included in this section, is of particular interest to Darwin, for it is a domesticated animal which at the same time is an example of "Darwinian atavism," through which a domesticated variety regains lost ancestral traits as a result of human neglect. This "indigenous variety" of the pampas stock is an example of "involution," which Sarmiento blames on the natives' indolence. This is why the *ñata* cow, though of significant scientific value, has little value at all on the international market. Sarmiento transforms a scientific theory into a matter of political economy and begins to foresee its possible applications. For Sarmiento, there is a need to "civilize" the cattle. By this he means that the Argentine countryside should be modernized, so that Argentina might better position its products on the world market.

This is no small matter in a country of stockbreeders. According to Sarmiento, it was from local stockbreeders that Darwin took valuable ideas about the "art of *varying species*. It was from them that Darwin took his first notions, right here, in our very pastures." It is true that techniques of artificial selection were not unknown to local stockbreeders: In 1813 a British stockbreeder by the name of Hasley introduced Merino sheep into Argentina, and a few years later, in the 1820s, President Rivadavia imported Merinos along with South Down sheep, which were crossed with the local breeds in order to improve their quality. The year 1820 was also the year the *toro tarquinus*, initiator of generations of Shorthorn cattle, was brought to Argentina. The first records detailing the pedigree of the Shorthorns mark the beginning of scientific stockbreeding in Argentina. This was later followed by books such as Daniel Pérez-Mendoza's *Tratado del ganado lanar* (*Treatise on Wool-Bearing Stock*, 1858), or, later still, Juan Harrat's *Estudios prácticos sobre la cría y el refinamiento del ganado lanar* (*Practical Studies into the Breeding and Refinement of Wool-Bearing Stock*, 1885).

THE WOODPECKER OF THE PAMPAS

(Jan 28th, 1870)

William Henry Hudson

A THIRD LETTER ON THE ORNITHOLOGY OF BUENOS AIRES, ADDRESSED TO THE SECRETARY BY MR. W. H. HUDSON:

There are four woodpeckers in the Province of Buenos Aires. Two of these, the *Picus mixtus* and the *Chrysoptilus chlorozostus*, you have seen in my collections; to both these birds the natives have given the vulgar name *Come-palo* or Wood-eater. Both of these species are quite common in the places they frequent, and are occasionally seen in the thickets South of the Rio Salado; but this is the extreme Southern limit of their range, and they prefer the Sayno forests bordering on the Rio de la Plata. The *Chrysoptilus chlorozostus* is sometimes seen to light on the ground, apparently to feed on worms and ants; its cries are when excited, loud, rapid and shrill, at other times it modulates them to notes exceedingly soft and sorrowful.

The third species, the *Carpintero Blanco* (White Carpenter-bird) affords another illustration of the influence of the riverine wood in introducing new species from the North to this country: for this bird, which is a native of the Northern States of La Plata, is occasionally found within a few miles of the city of Buenos Aires, though never, to my knowledge, South of it. Probably the divergence from the typical mottled color of the woodpeckers is greater in this species than in any other. I am not acquainted with its habits.

The fourth species is the *Carpintero*, more widely distributed, and better known than the other members of the genus to which it belongs, and also of great interest in reference to the erroneous account of its habits in Mr. Darwin's work, which makes it worthy of particular attention.

However close an observer a naturalist may be, it is not possible for him to know much of a species from seeing perhaps one or two individuals, in the course of a rapid ride across the pampas. Certainly, if Mr. Darwin had truly known the habits of the bird, he would not have attempted to adduce from it an argument in favor of his theory of the Origin of Species. In Chapter VI of his well-known work on the subject, the author speaks of the altered habits caused by change of habitat and other extraneous circumstances, and infers that it would be an easy matter for natural selection to step in, and alter an animal's structure so as to make a new species of it, after its habits have been so altered. He thus asked whether "there can be a more striking instance of adaptation given than of a woodpecker for climbing trees and for seizing the insects in the chinks of the bark?" and which is consequently a "woodpecker which never climbs a tree" (*Origin of Species*, 4th ed., vi, pp. 212–213).

The perusal of the passage quoted by one acquainted with the bird referred to, and its habitat might induce him to believe that the author purposely wrested the truths in order to prove his theory; but as Mr. Darwin's "Researches" were written before the theory of Natural Selection was conceived, and abound in similar misstatements when treating of this country, the error must be attributed to other causes. The facts are, that besides orchards, and groves of willow, poplar, etc. which have been planted wherever the plains are settled, there is also the continuous wood, which I have already described, growing on the shores of the La Plata.

South of Salado river, the numbers of wild trees have given a name to a large department of this Province. There is also in the vicinity of Dolores, 150 miles South of Buenos Aires city, a very extensive forest. All these woods are frequented by the Carpintero, where he may be observed climbing the trees, resting on his stiff and frayed tail feathers, and boring the bark with his bill as other woodpeckers do. But his favorite resort is to the solitary ombú, a tree found over a great extent of the plains of Buenos Aires. This tree attains a considerable size; there is one situated within fifty paces of the room I am writing in that has a trunk which measures at a height of 3 feet above the ground, 30 feet in circumference. This very tree was for years a breeding place for several Carpinteros, and still exhibits on its trunk and larger branches scars of old wounds inflicted by their bills. The wood of the ombú is very soft; and the Carpintero invariably bores for breeding where it is green and sound: the hole it forms runs horizontally about 9 inches into the tree, then slants upward a few inches more, and at the end of this passage a round chamber is excavated to receive the eggs.

The Carpintero frequently lights on the ground, where it is seen to feed on ants and larvae, and is sometimes found several miles distant from any trees. This, however, is very rare; and it is on such occasions always apparently on its way to some tree or trees in the distance. It very rarely takes a long flight, but travels by very easy stages. These circumstances have led to its being described as living exclusively on the ground. Outlying the regions abounding in trees, and which I have described as the habitat of the Carpintero, there are vast tracts in the Southern and Western portions of Buenos Aires, where, in truth, "not a tree grows"; but in these regions the Carpintero is never seen. It is not only the erroneous account of the bird's habits that makes Mr. Darwin's mention of it peculiarly unfortunate, but also because this bird rather affords an argument against the truth of Mr. Darwin's hypothesis. Mr. Darwin describes it as a perfect woodpecker, not only in conformation, but in its coloring, ondulatory flight and shrill of obstreperous cries. It is plain, then, that natural selection has left it unaltered; and is it not reasonable to suppose that, if there was such an agent in nature, it would have done something to alter this species, placed as it is in a situation so badly adapted for its structure and habits? But, in truth, natural selection has done absolutely nothing for our woodpecker. Its colors are not dimmed, nor its loud notes subdued; but even when it traverses the open country it calls about it the enemies from which it has little chance to escape. Natural selection not has endowed it, for its safety, with the instinct of concealment, so common in the true pampa birds. Its peculiar flight also, so admirably adapted for gliding through the forest, here only excites the rapacious birds to pursuit. In fact, the residence of this species in a region of which the conditions seem inimical to its preservation, so far from modifying, seems rather to have intensified its characteristics.

Compared with the other woodpeckers of this portion of South America, in structure, size, color, voice, and flight, it is the type of the genus. The habit of occasionally perching on the ground it possesses in common with other species, but it never roosts on the ground, like the true pampa birds, never builds a nest or burrows in banks, like the Patagonian Parrot, nor ventures on to those vast and treeless plains that border on its habitat. Scarcity of provisions, and seeking for trees better adapted for breeding, with, perhaps, other reasons, have probably led to the distribution of this species over a great extent of country.

Twenty years ago, which is as far back as my recollection extends, the Carpintero was rather a common bird, but it has now become so very rare, that for the last four years I have met with only three individuals.

THE RESPONSE

(1870)

Charles Darwin

In the last of Mr. Hudson's valuable articles on the ornithology of Buenos Ayres, he remarks, with respect to my observations on the *Colaptes campestris*, that it is not possible for a naturalist "to know much of a species from seeing perhaps one or two individuals in the course of a rapid ride across the pampas." My observations were made in Banda Oriental, on the northern bank of the Plata, where, thirty-seven years ago, this bird was common; and during my successive visits, especially near Maldonado, I repeatedly saw many specimens living on the open and undulating plains, at the distance of many miles from a tree. I was confirmed in my belief, that these birds do not frequent trees, by the beaks of some which I shot being muddy, by their tails being but little abraded, and by their alighting on posts or branches of trees (where such grew) horizontally and crosswise, in the manner of ordinary birds, though, as I have stated, they sometimes alighted vertically. When I wrote these notes, I knew nothing of the works of Azara, who lived for many years in Paraguay, and is generally esteemed as an accurate observer.

Now Azara calls this bird the Woodpecker of the plains, and remarks that the name is highly appropriate; for, as he asserts, it never visits woods, or climbs up trees, or searches for insects under the bark. He describes its manner of feeding on the open ground, and of alighting, sometimes horizontally and sometimes vertically, on trunks, rocks, &c., exactly as I have done.

He states that the legs are longer than those of other species of Woodpeckers. The beak, however, is not so straight and strong, nor the tail-feathers so stiff, as in the typical members of the group. Therefore this species appears to have been to a slight extent modified, in accordance with its less arboreal habits. Azara further states that it builds its nest in holes, excavated in old mud walls or in the banks of streams. I may add that the *Colaptes pitius*, which in Chile represents the pampas species, likewise frequents dry stony hills, where only a few bushes or trees grow, and may be continually seen feeding on the ground. According to Molina, this Colaptes also builds its nest in holes and banks.

Mr. Hudson, on the other hand, states that near Buenos Ayres, where there are some woods, the *Colaptes campestris* climbs trees and bores into the bark like other Woodpeckers. He says, "it is sometimes found several miles distant from any trees. This, however, is rare, and it is on such occasions always apparently on its way to some tree in the distance. It here builds its nest in holes in trees." I have not the least doubt that Mr. Hudson's account is perfectly accurate, and that I have committed an error in stating that this species never climbs trees. But is it not possible that this bird may have somewhat different habits in different districts, and that I may not be quite so inaccurate as Mr. Hudson supposes? I cannot doubt, from what I saw in Banda Oriental, that this species there habitually frequents the open plains, and lives exclusively on the food thus obtained. Still less can I doubt the account given by Azara of its general habits of life, and of its manner of nidification. Finally, I trust that Mr. Hudson is mistaken when he says that any one acquainted with the habits of this bird might be induced to believe that I "had purposely wrested the truth in order to prove" my theory. He exonerates me from this charge; but I should be loath to think that there are many naturalists who, without any evidence, would accuse a fellow worker of telling a deliberate falsehood to prove his theory.

THE MUÑI-FELIS BONAERENSIS (1845)

Francisco Javier Muñiz

Dear Editor of the *Gaceta Mercantil*:

Having produced a brief report about an extraordinary fossilized quadruped found in the great alluvial layer that covers our province just below its surface,[1] and which extends across the greater part of Argentina's territory, I propose that it be published, if it be to your liking, in your fine periodical, as I find the discovery that has motivated it to be worthy of the public light. Since the laudable *Mr. Darwin* has not made known the species about which I plan to give notice, after his interesting papers on the Patagonian Coast and other parts of the Republic from 1832 to 1836, I am certain that I am the first to offer it, in the following succinct description, to the examination of the scholars who dedicate their lives to the study of these unimpeachable witnesses, victims of both frightening and devastating catastrophes.

In the skeleton or fossil that I mean to discuss, I recognize an individual of the genus *Felis*, that, based on certain structural peculiarities, resembles the lion. If anything distinguishes it from this formidable dictator of the quadrupedal tribes, living or fossil, it is precisely that it must have been, within its very genus, even fiercer and more destructive, stronger and more fearsome for the animals that make up those tribes.

Of necessity I have had to give the skeleton a name, which will be rectified, if it be inappropriate, by the naturalists who are well educated in the laws governing the structure of animals, in their distinctive traits or in the identity of their specific or primitive forms. This name is—in recognition of the place in which it was found—*incisoriis aduncis muronatis* due to the sharp, hook-like shape of its incisors, a trait which could lead one to believe that it belongs to a different genus that the one to which I have directed it. But if such a modification in those teeth and one organic oddity or another constitute a new species in geology and for zoology; if it is unheard of in the order Carnivora, whose dental number and configuration is nearly invariable; and this is even truer in the genus *Felis*, in which only the two most external fangs of the upper mandible are pointed, the exception in this case only matters, osteologically speaking, in terms of a heightened ferocity, a devastating demeanor on an inconceivable level toward the rest of its currently recorded congeners.

As to the rest, as shall be seen later on, its generic identity is confirmed by the coexistence, placement, and shape of its other organs, particularly the molars, a shape that is clearly recognizable as the basic trait of this group, by which one may judge whether one of its species is more or less ferocious in nature, ranking from the cat to the bear.

ESSENTIAL OSTEOLOGICAL TRAITS

As a carnivore, it exhibits outwardly convexed zygomatic arches; orbits incomplete in the rear and confused with the temporal fossa, transverse fossa, and a cylindrical portion of the mandibular condyle; a deep glenoid cavity; in the posterior edge of that bone, beneath the condyle, a deep semicircular groove; the scant coracoid process; the spine of the scapula that ends in a large acromion; the irregular ellipse on the superior end of the radius; the oval shape of the articular surface of its inferior end; the hooked process on this part and its deep grooves which imply the strength and direction of the digital tendons, etc.

As a member of the genus *Felis*, the facial outline is short and broad, due to the transverse width of the bones that form the upper part of the nasal opening; the incisors are grooved on their inferior surface; the molars have a triangular, sharp, serrated shape; the mastoid process is much larger than the occipital condyle; the curvature of the medial border of the scapula; the grooves on the lower edge of the acromion; the deep cavity in the olecranon, indicative of ample support for running; the considerable length, width, and roughness of this apophysis; the structure and disposition of the bones in the carpus, metacarpus, and particularly the peculiarities in the penultimate and terminal phalanges in the five digits armed with large nails that come to a fine point, are retractable, sharp-edged, and hooked, and are set in an extremely durable sheath.

In the rear limbs we find the conspicuous narrowness of the pelvis and the parallelism of the spine and the end of the ilium; the head of the femur below the level of the greater trochanter and the former's almost non-existent neck; the inferior extremity of this bone which is longer than its anterior-posterior diameter with respect to its width; the tibia, a difficult bone to discern in quadrupeds, which exhibits a rounded intercondylar eminence; consonance of the bones of the tarsus and the foot as well as the claws, as with the hands. There you have some bits of evidence that the animal to whom this skeleton belonged was a member of the genus *Felis*.

The structural similarities between this animal and the lion consist of the considerable concavity or fossa situated between the postorbital processes (a distinctive feature of the species *leo*); the near parallelism of the profile of the face with that of the head;[2] the superior articular surface of the radius shaped like the concave portion of a ring and whose inferior end is divided into two processes by a deep groove; and the ulnar cavity which is deep and occupies the middle part of the trochlea. There are, moreover, further attributes common to both species.

However, the difference between them is shown by the profile of the cranium, which follows a somewhat straighter line in this skeleton than in the *lion*; the greater space in the former between the postorbital processes and the occipital spine; the length of the sagittal crest which is longer, even more so than in the *Felis spœlea*;[3] the flatter and more horizontal forehead than in this feline, the tiger, and the jaguar; the inferior part of the zygomatic arch which is more closed-in and semicircular than in all of these; its larger infra-orbital foramen.

What most characterizes this new and previously unknown species, and sets it apart from all the others included up till now in the genus *Felis*, is the shape and large size of its canines and the lack of two false molars in the upper jaw and in the lower one as well. Feline animals have twelve incisors, four canines, and fourteen molars—thirty in all. This skeleton has only twenty-six: twelve incisors, four canines, and ten molars, the same number as a cat's set of *deciduous teeth*, but in this case it is rather its set of *permanent teeth*.

In spite of this defect and some other peculiarity here or there, which one would cite in a complete description of this fossil, the fossil is still a member of the felines. The lack of one or more teeth and the distinct volume of its limbs do not constitute a generic difference, when moreover it is the case that there exist invariable traits in structure, osteological relationships, and uniformity of the joints, especially correspondence in the molars, whose shape never changes at all in species of the same genus, and in the hands and claws—an identifier which, being the same in the skeleton and the genus *Felis*, clearly indicates a common system connected to predation, and the same propensity to feed on live prey.

[. . .]

Table 1.

	Bonaerensis head			*Felis spoelea*			*Lion*		
	Feet	*Inches*	*Lines*	*Feet*	*Inches*	*Lines*	*Feet*	*Inches*	*Lines*
From the border of the incisors to the occipital foramen	1	2		1	1		1	1	3
From this same border to the occipital spine	1	4	3	1	4	2	1	3	7
From the middle of the frontal bone's postorbital processes to the basilar edge of the occipital foramen	1	1	12		8			7	

I will now venture a few osteological measurements so that they may serve as a comparison with other species of the same genus; and give an idea of the size, strength, and might of that erstwhile destroyer of the animal life of this region.

It looks, then, as though the skeleton has been shown to belong to an enormous species, larger than any other of the genus *Felis*. The *Felis spoelea*, which could only sustain a comparison with it in certain of its parts, and even exceed it in some of them, succumbs before a comparison with these essential characteristics. The remains that are believed to belong to this species are more or less, up till now, the head that I have used as a means of comparison, two fairly

Table 2.

Tiger	*Feet*	*Inches*	*Lines*
From muzzle to nape, according to Mr. Azara (the largest dimensions with which I am familiar)	1	$^1/_2$	
According to the learned Count Cuvier		10	

Table 3.

Muñi-felis (other measurements)	*Feet*	*Inches*	*Lines*
From the anterior alveolar border of the incisors to the basilar edge of the occipital foramen (the cranium contains the largest measurement in the skeleton)	1	10	
Height from this same basilar edge to the occipital crest		5½	
From the crest to the anterior border of the nasal bones		11	12
From the basilar edge of the occipital foramen to the posterior border of the vault of the palate		7	4
Distance between the furthest-protruding parts of the zygomatic arches		9	12
Width of the extremity of the nasal bones		3	
The same of the skull from orbit to orbit		5	4
Longitudinal diameter of the occipital foramen		1½	
The same, transversally		1½	
Length of the mandible from the posterior angle to the upper anterior alveolar border of the incisors		10½	
Height of the ramus of the mandible from its angle to the highest part of its head		2½	
The same from the inferior border of the angle to the highest part of the coronoid process		4	
Total length of the upper fangs measured in a straight line		10	3

large bones from the metacarpus, and a radius which, though slightly longer than that of this skeleton, is smaller in any case in other dimensions. The fossil fang (supposed to be from the same species) which naturalists have agreed is the largest of all, and which they consider to be the rarest specimen of its kind, was unearthed on Hauteville Street in Paris at a depth of twenty feet among other bones of horses, and in its enameled portion alone measures two inches three lines (0.055). The difference could not be more marked with those of this skeleton, and nevertheless the great Mr. Cuvier says that there is no comparison with even the largest lion and tiger skulls. With what, in this case, could the fangs of our fossil be compared, which are so much larger than even those of Mr. Goldfuss's great *Felis spoelea*? And it appears certain that the skeleton (probably of a female) did not belong to one of the largest animals of its species, as a calcaneus found in one excavation indicates that there were even stouter ones.

Although nothing can be said about this great cat of our plains which is not highly hypothetical, it would appear, in any case, that its form was more graceful (if the mind can accept this pleasant quality in something that horrifies and frightens us) than that of the jaguar, which is considered as having less repulsive proportions than the chiefs of the dark and treacherous feline family, the lion and the tiger. Its longer head, the longitudinal dimensions of its limbs, the dimension of the spinal column, the positioning of the pelvis, the length and convexity of the ribs, lead one to surmise that this terrible beast of our ancient and solitary territories had a comparatively superior physique.

As for the solidity of its bones, particularly those that most require this quality, such as the teeth, the ones at the extremes are remarkably compact. The teeth, the ribs, the bones of the hands and feet, all make a distinctly metallic sound when struck together.

The strength of the structure as a whole leaves no room for comparison, not even with the large and ferocious African lion, whose neck it would sever and whose most hidden entrails it would uncover with one bite from its fangs. The remarkable robustness and magnitude of its limbs, the roughness of the strong tendon insertions that cover them, the humeral crests to which the ligaments of the muscles of movement in the hand and its digits attach, those of the radius to which the ligaments used in pronation and supination attach, the firmness, in short, of the brachial joints, the wide chest, the irresistible force of the five unguiculate digits which are unrivalled in their genus, and of its uncinate or short incisors, and the still more terrifying apparatus of fangs with two cutting edges of a rare and particularly large size; these are all unequivocally signs pointing to the muscular and terrifying king of the jungle, to the chief monster of the feline tribe, the most indomitable and indefatigable of hunters, the boldest and most ferocious in its crude and insidious attacks.

If it is true what they say, that the lion is the enemy of the elephant; if the *Megalonix* had been the enemy of the fossil animal found in Ohio, as President Jefferson mistakenly supposed; the *Bonaerensis with hooked and sharpened incisors* would probably have been the implacable pursuer of the heavy *Megatherium*; of the immensely bulky *Mastodon*, of the horse, and even of the great plated *Giant Armadillo*, along with other peaceful quadrupeds who lived, at one time, in these regions—and it would be thus, because while it was not endowed with a tall stature or large mass, it was agile and strong, and more than sufficiently ferocious and astute, as its constitution indicates.

Though the opinion of the past century's great naturalist was erroneous regarding the smallness that he attributes to today's American fauna when compared with their European counterparts, it is true, nonetheless, that in terms of the animals that populated it after the last catastrophe, the New Continent cannot oppose any animal larger than the fierce *jaguar*, the *tapir* with its surprising shape, or the bloodthirsty *puma* (*Felis discolor*[4]),[5] to the other hemisphere's enormous pachyderms–*the elephant, the hippopotamus, and the rhinoceros.* Yet, in compensation, we find, like debris scattered by some great shipwreck, in the south as well as the north of the New World, colossal skeletons of herbivores and even of amphibians that are unrivalled on either the Old Continent or the Islands. In fact, there is not a living quadruped there, nor has any fossil been seen (if we exclude some oviparous animals that are only larger in length such as the *Megalosaurus* and the *Ichthyosaurus*) which could compare with the *thirty foot* long and *fifteen foot* tall *Missourium*; nor do we find in the existing or fossil species of the *Felis* genus any which in size and fierceness exceeds the *Bonaerensis*, in comparison with which the greatest *Felis* of the caverns (*Felis spoelea*) would not amount to much, nor the smaller one (*Felis antiqua*), nor the large or small species of *Felis outside the caves.*

To close this report, I must openly confess that, as I am cut off from all relations that could enlighten me, and lack all instruction necessary to make a description of this nature conform to proper standards, I must implore the naturalists' indulgence in condoning my inaccuracies or errors. As I found the discovery of this new species to be interesting, and since there was no one to whom I could delegate the main details of its structure, I have had, against my will, to come face to face with intellectuals' judgment and criticism, revealing to them my ignorance in these matters. But necessity, dear Editor, exculpates me, as it is she who has placed me in this conflict and not any conceit of mine as to some false proficiency.

I kiss, kind Sir, your hands with great esteem and respect, your humble servant,

Francisco Javier Muñiz

Villa de Luján, July 1st, 1845. Year 36 of our freedom, 29 of our independence, and 16 of the Argentine Confederation.

P.S. If my dear Editor were to decide to admit into his illustrious periodical the preceding article, he may consider repeating this distinguished for the observations that I will send him in due time about the *Megatherium*. As the remains of this enormous quadruped from our pampas have garnered a well-deserved celebrity, it seems to me that any information which tends to shed light upon all things relating to its structure which have been omitted or insufficiently specified, or any useful remark, or anything which expands upon the known details of its history, already so extensive, would be well received by those who seek knowledge, or find worthy of appreciation this sort of antiquities.

NOTES

1. OAN: Though the depth is variable (due to accidental features of the terrain) at which we find the remains of diverse species of extinct animals scattered across the province, we may adopt nine to ten feet as a median depth, with respect to the upper part.

2. OAN: These noticeably unequal lines in the true tiger are even more so in our *jaguar*, whose skull is much more prominent than those of the great cats of Africa.

3. OAN: For these comparisons we have made use of the plates in the great work by the renowned M. Cuvier, *Sur les ossements fosiles*.

4. TN: Commonly classified as *Felis concolor*, the same as the North American puma, cougar, or mountain lion.

5. OAN: The lion of our plains. *Yaguarí* among the Guarani. *Onça parda* among the Brazilians.

A LETTER TO MUÑIZ

(1847)

Charles Darwin

Down, Farnborough, Kent,
February 26, 1847.
DR. F. J. MUÑIZ, Buenos Aires.
RESPECTED SIR,

Your letter of August 30, with the papers which you were so good as to send me, reached me only a short time since, owing to the protracted illness and absence from London of Mr. Morris, through whom they were sent. I have lately heard from Mr. Morris that you wish to dispose of your fossil remains on some pecuniary arrangement, which I did not fully understand from your own letter to me. I have given Mr. Morris my opinion on this head, so will not here repeat it; but will only say that I conceive the only feasible plan would be to send your fossils here to some agent to dispose of them. No society will purchase anything of the kind without having them inspected, and most societies only receive presents. Your specimen of the Muñi-felis must be a noble one; I suspect it will turn out to be a *Machairodus*, of which there are some fragments in the British Museum from the pampas. I will endeavour to get your paper translated and inserted in some scientific periodical. Your account of the earthquake in the pampas has surprised me; I never heard of one in any part further east of the Cordillera than at Cordoba. If you will inform me whether you read English I shall be happy to send you a copy (if you will point out some channel) of my "Geological Observations on South America," lately published; I do not think it worth sending them without knowing whether you read English, which I fear is not probable. Your pamphlet on the scarlet fever I will present to the Royal College of Surgeons.

I cannot adequately say how much I admire your continued zeal, situated as you are without means of pursuing your scientific studies and without people to sympathise with you, for the advancement of natural history; I trust that the pleasure of your pursuits affords you some reward for your exertions. Some time since you were so kind as to send me through Mr. E. Lumb some most curious, and to me most valuable, information regarding the *ñata* oxen. I should be deeply obliged by any further facts about any of the domestic animals of La Plata; on the origin of any "breed" of poultry, pigs, dogs, cattle, etc. I should be much interested by a brief description of the habits and appearance of the pigs, dogs, etc., which have run wild, and especially on the habits of these wild breeds, when their young are caught and reared. Will a puppy of one of the run-wild dogs, if brought up carefully, be as tame as a common dog? Any information on all such points would be of real service to me; and my address, should you find time to write to me, will always be that at the head of this letter. I most sincerely wish you all success in your admirable labours, and if at any time I can be of any service, I shall be happy to be so; but I am sorry to say I am not connected with any mercantile establishment and cannot recommend agents, etc., etc.

With much respect, I beg to remain, Sir, Your obliged and obedient Servant, Charles Darwin.

P.S.—I omitted to state that Prof. Owen has heard that a collection of bones from Buenos Aires some time since arrived at Paris.

ÑATA OXEN

(1885)

Domingo F. Sarmiento

Such is the odd name that Darwin gave to a variety of cow, which originated and apparently became established in Buenos Aires, that he learned of through a special study sent to him by Dr. Muñiz, a draft of which is to be found among Dr. Muñiz's papers under the name, *"Response to the seven questions concerning the* ñata *cow asked of the undersigned by Mr. Enrique Lumb."*

The theory of the formation of species through natural selection must have already been incubating in the mind of this bold innovator at this time, since the *Voyage of the Beagle* states that during his visit to these countries, Patagonia, the Galapagos Islands, etc., he had his first intuitions. "*Most curious*, and to me *most valuable*," says Darwin about the study of the *ñata* cow, and to this end he requests that Muñiz inform him of "any further facts about any of the *domestic* animals of La Plata; on the origin of any 'breed' of poultry, pigs, dogs, cattle, etc. I should be much interested by a brief description of the habits and appearance of the pigs, dogs, etc., which have *run wild*, and especially on the habits of these wild breeds, when their young are caught and reared. Will a puppy of one of the run-wild dogs, if brought up carefully, be as tame as a common dog?"

It would be of little interest today to read that series of questions about the existence and later extinction of a variety of cow that was bred on the ranches of Buenos Aires, were it not related to the theory of evolution that later gained such celebrity, and if the records of Dr. Muñiz did not contain several references to it, in addition to the part of said records that Darwin refers to and cites in his *Voyage of the Beagle*.

Ñata cows were introduced to the ranches by Indians, who brought them in exchange for the goods they came to stock up on in Buenos Aires. "Before the revolution," states Dr. Muñiz, "it was the Christians who frequented the Indian camps in times of peace. It was forbidden for infidels to cross over the border into the interior without certain restrictions, which, although in and of themselves rather simple, must have been even more of a deterrent for the man of nature than the tribute and border patrols would have been for the commerce of civilized men."

In addition to the blankets, frieze, ostrich feathers, reins, colt-skin boots, salt, belts, cloths, etc., which the Indians exchanged for tobacco, whisky, baize, spurs, bridles and other riding gear, knives, etc., they also brought cattle. Seldom small or young, generally large and fat as the traders demanded. By this means, the *ñata* cattle, which made up, according to the testimonies of the old ranchers of the Province (negotiators with the barbarians), a large portion, if not the largest, of their stockyards, was first introduced into the *partidos*[1] that had more trading contact with the natives. And so it was that from Pergamino, Rojas, Areco, Guardia de Luján and Navarro, *ñata* cattle spread to the south, north, and interior of the plains of Buenos Aires.

We prefer the citation Darwin makes in *The Voyage of the Beagle*, in his own words, of the description of the *ñata* cow, citing Muñiz and adopting his ideas, since in his pen they carry the seal of scientific acceptance.

"On two occasions," he says, "I met with in this province some oxen of a very curious breed, called ñata or niata. They appear externally to hold nearly the same relation to other cattle, which bull or pug dogs do to other dogs. Their forehead is very short and broad, with the nasal end turned up, and the upper lip much drawn back; their lower jaws project beyond the upper, and have a corresponding upward curve; hence their teeth are always exposed. Their nostrils are seated high up and are very open; their eyes project outwards. When walking they carry their heads low, on a short neck; and their hinder legs are rather longer compared with the front legs than is usual. . . . Don F. Muñiz, of Luxan, has kindly collected for me all the information which he could respecting this breed. From his account it seems that about eighty or ninety years ago, they were rare and kept as curiosities at Buenos Ayres. The breed is universally believed to have originated amongst the Indians southward of the Plata; and that it was with them the commonest kind. Even to this day, those reared in the provinces near the Plata show their less civilized origin, in being fiercer than common cattle, and in the cow easily deserting her first calf, if visited too often or molested. It is a singular fact that an almost similar structure to the abnormal one of the *ñata* breed, characterizes, as I am informed by Dr. Falconer, that great extinct ruminant of India, the Sivatherium." The whole of the foregoing text is taken word for word from the questions and responses given by Muñiz to Lumb.

In fact, these general traits bring to the mind of the naturalist or simple traveler the image of the bison, who writers on the artificial breeding of cattle hold to be an ancestor of our European cow. Darwin has applied the term atavism to a tendency contrary to that which produces variability among the individuals of a species, which from time to time makes primitive ancestral forms reappear, as in the horizontal bands on the legs of colts in the Argentine pampas that he believed were traces of the zebra, ancestors, according to him, of the horse. The appearance of a type of domesticated cattle in these very pampas with a shorter neck, with a more prominent nose, with a lower head than that of European cattle, would lead one to the theory of atavism, neglect, much of the perfection of plant and flower breeds is obtained through meticulous cultivation and constant diligence in breeding the most perfect varieties.

In the case of the *ñata* cow, which degraded the stock of the Buenos Aires plains to the point that *ñatas* came to form the greater part of the cattle, there is not a point nor a factor shrouded in doubt or darkness. Cattle were introduced to America[2] by the Spanish conquistadors. The stock, by and large red-brown, as evidenced by the predominant color among its present-day descendents, was of Dutch origin, according to the opinions of Mr. Leonardo Pereira. Its aptitude for producing milk would support this conjecture. Since today we have pure-bred Dutch stock at hand, we can see for ourselves that all of the deficiencies of our Argentine cattle are gradually acquired degenerations, the product of having neglected the cattle and given reign to their instincts, out on the immense expanses of the pampas with neither fence nor pen. When visiting the cattle market founded in 1867 on the outskirts of Chicago, the cattle men asked me to describe for them the cattle of the pampas. No easy task for someone who understands little to none in matters of breeding; but with some effort I began to say, "huge head, large horns, long legs, prominent bones . . . like that one over there," I interrupted myself, "like that one over there, and that other one there. . . ." The general laughter confirmed the accuracy of my description. *Spanish cattle*, shouted one of them. It was, in effect, a group of cattle from Texas, where they do not make butter and ranchers or gauchos handle the cattle on horseback with lassos, just as in the Republic of Argentina.

On top of the widespread degeneration of all the cattle left to their own devices across Spanish America, the Indians brought about a still greater degeneration in cows that were freed unto a life still wilder than that of the tame or native-born pampas cattle, so much so that they began to return to the characteristic form of the hump-backed bison. Without intending to, Dr. Muñiz puts forth the evidence. "Seventy or eighty years ago," he says in 1822, "it was extremely rare to find that variety on the ranches of Buenos Aires. Subsequently, when the relations between Christians and the Pampas and Ranquel Indians grew freer and safer, the barter trade facilitated the introduction of that variety of cattle." That it was a simple degeneration due to the savages' negligence is well attested to in Muñiz's narration. "It has been observed," he says, "as far as their nature is concerned, that they are unusually surly. The cow runs away and leaves her calf when a farmhand comes too close, and it is difficult to make her take it again. As the *ñata* calf is not as corpulent or strong as the common one, and having, rather, an unfavorable physiognomy and a sickly, almost rachitic appearance, it is widely considered in our country to be inferior to the common cow. As such, far from proliferating on its own, the breed is maintained by the particular tastes of one rancher or another. It is excluded from the market because of its defective leather, since this animal's head is so short that the leather comes out rounded and short in the muzzle, thus diminishing its value."

Another degeneration in the already slightly degenerated European cattle was the *mocha* cow, a contemporary of the *ñata*, "cattle similarly inferior to the common sort, since apart from lacking horns, which always entail a certain value, they are not useful as oxen, nor hardly as dairy cows either, since it is hard to handle them around the neck for these services."

"It is attested that on the pampas ranches of that period, these animals and the *ñatas* were *more numerous than the common variety*."

This assertion, twice repeated by such a discreet observer as Dr. Muñiz, who for so many years lived on the plains at the beginning of this century, and who refers to testimonies reaching back to seventy and eighty years prior, that is to say the beginning of the 18th century, prove to the point of obviousness that in the last century cattle raising in Buenos Aires descended into the lowest state of degradation and barbarity. The Indians brought in new degenerate breeds that harken back to the wild bull, the bison, the aurochs, much as wolf-like forms might reappear in the runaway dogs of the pampas, their ears perking up and sharpening, as, similarly, the stray dogs of Constantinople and Cairo take on the characteristics of the jackal, their supposed neighbor and progenitor. In this sense Dr. Muñiz's notes on this subject are very instructive; they were of much use to Darwin, and should be of even more use to us, as cattle raising constitutes our principal source of wealth.

Today it is more important than ever to call attention to the faulty practices in cattle raising that jeopardize its value, and will end up diminishing it completely if ranchers do not make haste to correct this degeneration through neglect of Spanish cattle, introduced to America four centuries ago, and left on their own without the meticulous care of man.

Because the sort of graminaceous vegetation growing on the pampas determined the development of cattle raising there, one can say that these are essentially cattle lands. And yet, since the production of cattle is greater than the relatively small population's demand for meat, it has been a failed business up to the present because the meat cannot be exported, save in extremely poor quantities and conditions. One could say that if there had been in some of the world's countries large-scale dog breeding, then Argentina's industry would have been their top provider of meat, since the jerked beef made in our salting rooms was eaten only by slaves in Brazil and Havana, and even they have begun to throw it by the wayside ever since the breezes of dignity picked up following slavery's abolition. Starting this year, 1885, all of the markets in the world have been opened up for refrigerated meats, which can be easily and effectively transported; but we now must face what the farmer has ignored up till now: that the cattle he breeds are degenerate, and lacking careful breeding they can hardly adapt to the production of meat for European markets. A cow is a set of bones and muscles preserved inside a sheet of leather. Exposure to a harsh climate and a life on the open plains makes the leather thick, hard, and solid. If it has the plains all to itself it will end up with a large and fortified skeleton. If it does not have a constant supply of food and exercise its muscles will be reduced to a minimum, just the opposite of what hard work brings about in oxen. Argentine cattle are not fit for exportation as foodstuff, since along with them thousands of tons of nearly worthless bones must be transported, accompanied by a relatively small amount of fibrous matter, considering that three full skeletons must accompany every 60 *arrobas* of meat. Poorly-used land, poorly-used pastures, exorbitant hire prices for transporting cargo only to find repulsion in the European markets, where our meats play the same role as the *ñata* cow in other times, and our jerked beef today, for the cultivated people of London's East End will not accept such meats, and due to its terribly low price the meat merchants set it aside for the working-class masses of the West End.

The same is the case with our Merino sheep, which by the way are not quite as wretched as our cattle, though they are still inferior to English, Continental, and Australian Merinos, the latter of which, despite having to make a journey three times as long, are still preferred over the Argentine variety.

An Argentine woman writing from London tells us: "The meat market is magnificent. We looked for our own Buenos Aires meat without finding any, which was lucky for the meat as it would not stand comparison with hardly any of the rest of them known there, as they have meats from many countries, and each shop shows the origin of its meat: United States, Australia, Continent, Russia, etc."

Nevertheless, we have the solution at hand, and all we need to do is apply it with assiduity and a resolute plan to ensure once and for all the productivity of the Argentine countryside. Raising *Durham cattle* instead of the *ñata* cow, the *mocha* cow and the degenerated Dutch cows, leaves the doors to the world markets wide open for us, and with preference for our product. Fortunately the herds and breeding places already exist, and of the quality and quantity required to begin, within just a few years' time, the necessary phasing-out of the old sort of cows, which are either degenerate or unfit for human consumption. Just as the *Rambouillet* sheep has, through a process of selection, tremendously increased its production of wool to the detriment of other products, the fleshy parts of the Durham cow are artificially enlarged to the detriment of bones, horns, head, these being reduced to a minimum. There is, in addition, one more reason to change our system of raising cattle, even after the cattle ranches have been fenced in, one which we find in Dr. Muñiz's study. "When, in the great droughts that occur in this Province," he says, "which were, in this century, the deadly drought of 1806, and those of the years 1830 and 1831, in which *more than two million* cattle died due to a total lack of both pasture and water, then the cattle make use of their lips to scrape clean, like a horse, the smallest little branches and any small piece of straw that, unappetizing and dirty as it may be, might provide them with some miserable refection."

The geologist Bravard, in explaining the formation of the pampas' landscape, attributes its form to the dry spells that have *ab initio* scorched this land, leaving deposits of dust brought in by the *Pampero* wind. This is demonstrated by the layer covering the fossilized skeletons, which rest in the same spot where they died, without fractures or signs of having been dragged or displaced, such that the female is near the male, pointing to death by inanition, which means a lack of water or food. The same phenomenon took place in Ceará in Brazil but a few years ago, and it will happen again here, and we will not be able to take out a loan on rain in the markets of London, to be paid off by our descendents. So it is possible, then, that all our cattle may die off in just one or two years of drought; and only the system of setting aside reserves of feed, a practice already in use by intelligent breeders of fine cattle, can save our collective fate from that *Dies irae*, which could reduce the entire cattle industry to only a handful of animals saved under extraordinary circumstances.

Still, in what reveals an aberration in taste, or an indifference regarding the degenerations that a return to life in the wild can bring about in domestic animals, a certain author cites in passing, as any other current event, "the importation of the *celebrated* breeds of pampas sheep." Well the farmers of that period celebrated some unworthy things indeed! We know that there was an abundance of puny cows, perhaps strewn about the landscape for the eye's amusement, the value of their leather or the quantity of their meats not amounting to much at all; yet we now hold a late celebration for the Merino, negrete, Rambouillet, blackhead, and Lincoln breeds, for their products which are works of intelligence and civilization, the rewards of hard work, and applicable to the needs of mankind, so that we may forget about those degenerations that were brought to us, to our own shame, by the savages, and which were taken on with neither discernment nor reason.

NOTES

1. TN: A *partido* is a territorial division in Argentina which is essentially equivalent to a county.
2. TN: In Latin America, the word "America" is used to refer to both American continents, not to the United States of America.

II

SCIENTIFIC FANTASIES

Darwin's Followers and Detractors in Argentina (Sarmiento, Ameghino, Moreno, Holmberg, Burmeister, Lugones, Estrada, Alberdi)

HOLMBERG'S *TWO PARTIES IN CONFLICT*

Two chapters of Holmberg's *Two Parties in Conflict* have been included in this section. As mentioned in the introduction, the novel transposes the atmosphere of the Oxford Debate to Argentina, where this time around the parties are the Rabianists and the transformists (or Darwinists). The characters in each of the factions, although not completely transparent or readily identifiable, are associated with the historical figures of Burmeister, Moreno, Ameghino, and Holmberg, and politicians like Sarmiento, Bartolomé Mitre (1821–1906), Valentín Alsina (1802–1869), and Nicolás Avellaneda (1837–1885) also participate. Furthermore, each faction represents a position which is not only scientific, but also political and philosophical. Because Argentineans associated evolution with both the modernization of political and social institutions and the secularization of the state, Darwinism was called the "ideology of Progress." The novel's Rabianists, on the contrary, represent the old regime, and their conservatism is more scientific and political than religious. As a whole, the authors in this second section were chosen because they all took part in one collective debate concerning Darwinism.

HOLMBERG AND MORENO: NATIONAL HERITAGE

The choice of texts in this section goes further than the controversy parodied by Holmberg, and reveals a whole array of underlying political, ethical, and economic questions that were affected by the idea of Darwinism in this period. Evolution and the idea of the struggle for life provided arguments for the legitimizing principles in the military campaigns of annihilation and exclusion of indigenous peoples from the modern nation (i.e., Julio A. Roca's campaign of 1880). With two distinct and sometimes opposing styles, Holmberg and Moreno debate the ethicality of the extermination of the indigenous peoples. Although both seem to agree that their disappearance will be an inevitable outcome of evolution, Holmberg uses irony to question the "victory of progress," and Moreno focuses on rescuing indigenous skulls and cultural artifacts, including records of their languages, in order to construct the nation's past and fill his own museum. Despite their contradictions, both men openly visualize and desire the modernization of the Argentine pampas and Patagonia, which for them means industry, and the total integration of their country into the world commerce market.

As a part of this drive toward modernization, and perhaps as a counterweight to it as well, the process of institutionalizing the national heritage was sped up, along with its study, exhibition, and display in museums, scientific observatories, and university departments. Paleontology placed the nation's explored territories and its territories yet to be explored within a much wider time frame. Geological eras, antediluvian fossils, and primitive peoples situated a whole swath of the national territory in human prehistory. Moreno, in his article included in this section, thought that Argentina's paleontology should reveal to humanity "the secret of [its] first homeland [. . .] the precise location of the encampments of [its] ancestors, who with time would go on to build entire nations." Moreno's beliefs concerning the origins of humanity were fundamentally mistaken insofar as he supported the polygenist theory, and held that the native peoples of South America had evolved from a different primate ancestor than that of the Old World humans. His assertion that "[m]onogenesis is fading, and polygenesis has an increasingly high probability of being correct in its calculations," was clearly on the wrong side of history.

However, at the time, Sarmiento—faced with the fossil material found by Moreno and evidence for different geological eras—agreed with contemporary naturalists about the possibility of setting the first chapter in the history of human evolution on Argentine soil. These "prehistoric worlds," right or wrong, would become part of the nation's heritage.

SARMIENTO AND SPENCER

Nineteenth-century evolutionists drew up unilinear frameworks to explain humanity's development through time, in which the evolutionary measure of modernity was progress. These frameworks consisted of a succession of uninterrupted states with no gaps between them, wherein the scientist's task was to go about ranking the world's human societies in accordance with their level of development along this scale. In this way both the contemporary Tasmanians and the primitive pre-Columbians of the Americas could be situated together at the beginning of this series of evolutionary states.

Although Sarmiento declares himself a "Darwinist," and praises Darwin as the most prominent scientist in the history of humanity, he also states that he understands evolutionism in the vein of the social Darwinist Herbert Spencer (1820–1903), who gave him "a principle that is both beautiful and harmonious, in order to dispel doubt, which is the torment of the soul." And so evolutionism, for Sarmiento, would fill the void left after divine creation had been discredited, and would explain historical change and progress through different stages and figures: Classical Greece, the Renaissance with Copernicus and Galileo, the Protestant reform, Bacon, Benjamin Franklin, and on down to Darwin.

Most authors in this section completely disregarded the Bible; they did not hold creationism to be the logical alternative to evolutionism. For Sarmiento, evolution meant progress toward a civilized society, a goal that could be achieved through liberal education, to which a "barbarian" population, be it native or immigrant, could adapt and thus be transformed.

ALBERDI ON EDUCATION AND APPLIED EVOLUTIONISM

The selection included in this section from Juan Bautista Alberdi's book, *Luz del Día's Pilgrimage*, is a satirical attack on Sarmiento's confidence in the power of education to transform the species *Homo sapiens*. In this tale, a Darwinist Don Quixote (Sarmiento) sets out to "evolve" a multitude of sheep (the masses) into human citizens, a scathing portrayal of Sarmiento's application of Lamarckian theories of moral inheritance to the realm of education. In Alberdi's view, the fittest of the human races was the Northern European variety, and it was therefore the best population to target for immigration to Argentina. As we will see in more depth in part III, the controversy on immigration was also in many ways an offshoot of the debate surrounding Darwinism.

BURMEISTER: THE MODERNIZATION OF SCIENCE AND THE IDEA OF FIXED SPECIES

As part of the 1870s movement to institutionalize the sciences in Argentina, Sarmiento invited the distinguished Hermann Burmeister to give worldwide status to the nation's sciences by leading the museum of natural science. The institutionalization of science was a key task which needed support from international players, as the enlightened ideals of the time saw science as a central part of a sovereign and civilized nation.

The German national arrived in 1873, originally to take charge of the Faculty of Physio-Mathematical Sciences at the University of Córdoba. Two years prior the National Observatory had been inaugurated in the same city by Benjamin Gould (1824–1896), one year prior, in July of 1872, the Argentine Scientific Society had been created, and in the following year, 1874, the University of Buenos Aires was recognized and had Eduardo Holmberg and José María Ramos Mejía among its students.

Burmeister was the most famous anti-Darwinist in Argentina. In *History of Creation* (1843), the work with which he had gained an international reputation before his arrival in Argentina, Burmeister declared that he did not give any credit to Darwinism and stated that he did not believe in its "hypothetical conceptions without demonstrative power." For Burmeister, to accept Darwinism meant to open the flood gates of an epistemological revolution, to bring about a paradigm shift in the sciences over which, up to that time, he had held considerable authority.

History of Creation espouses the beliefs that all the species are fixed, and that they are all imperfect realizations of one original ideal, an ontology that stands in clear opposition to Darwin's idea of the species' random variability. Burmeister held that this conception of fixed species modeled after one ideal was especially true of the genus *Homo*. In his theory there was an underlying, perfect human ideal that was altered by the influence of the different habitats in which each race was conceived. Thus, while some contend that Burmeister is a racist, others hold that, despite his opinion that there are innate differences between the races, the theory that they are all imperfect copies of one perfect ideal implies that no race is better than any other. His idealism in this respect contrasts with his strict empiricism in rejecting the theory of evolution for its lack of empirical evidence. For Burmeister, evolution was "outside the realm of just experimentation." In any case, the same can just as easily be said of his ideal fixed species.

AMEGHINO THE DARWINIST

Aside from Holmberg, the strongest opposition to Burmeister's authority was Florentino Ameghino. In his day and age, Ameghino was not the most likely candidate to become an eminent collector of specimens and a renowned naturalist: He was born into a family of lower-class Italian immigrants, and, moreover, grew up outside the capital city, in the province of Buenos Aires. Ameghino was a radical Darwinist, as can clearly be seen in the text that appears in this section. The naturalist foresaw that, following the principles of Darwinian theory, a mathematical method could be devised to predict the paths taken by evolution, and thus reconstruct the missing links between known organisms. Because of his faith in the field's potential for mathematical precision, Ameghino ranked Darwinism among the exact sciences. It is worth pointing out that today his modest proposal to make a place for mathematics and prediction in naturalism pales in comparison with the advances made in the study of evolution since the discovery of DNA and genetics.

Like Moreno, Ameghino believed that human beings originated in Argentina. However, unlike his colleague and rival, he did not believe that just one race had originated there, but rather the whole of humankind. Still, as brilliant as he was in so many respects, Ameghino was mistaken on this point.

LUGONES ON AMEGHINO AND DARWINISM

Both Moreno and Ameghino form part of the Argentine national pantheon. Ameghino in particular was a paradigmatic figure of the liberal enterprise, who embodied the ideals of progress and socialism for his followers in Argentina. In this vein, Lugones wrote *Tribute to Ameghino* (1915), which is a biography of the scientist and an explanatory treatise on his theories. Lugones ushered in a trend of writing epic biographies that portray the nation's scientists as dignified role models, for example: *Ameghino (una vida heroica)* (*Ameghino (A Heroic Life)*, 1951) by Fernando Márquez Miranda; *El perito Moreno: Centinela de la Patagonia* (*The Expert Moreno: Sentinel of Patagonia*, 1949) by Carlos A. Bertameu; or *Francisco P. Moreno: Arquetipo de argentinidad* (*Francisco P. Moreno: Archetype of Argentineness*, 1954) by A. D. Ygobone. Of these biographies, the one that stands out the most is José Ingenieros's exhaustive study, *Las doctrinas de Ameghino* (*The Doctrines of Ameghino*, 1919).

Lugones was a Darwinist himself, but in his short story *Yzur*, he resorts to fantasy to explore the period's common and contradictory mixture of spiritualism, mesmerism, and science. Lugones's *Tribute to Ameghino* provides a clue to understanding the connection between the theories of Darwin and Ameghino. In this book he rejects, alongside Ameghino, all systems of classification that presuppose a hierarchical distribution of living things according to arbitrary concepts of superiority and inferiority. As he puts it, "[a]ll of the zoological branches come from, in this manner, one primitive organic trunk, and never one from another, as Darwinist transformism believed; on the contrary, they evolve in progressive divergence from one another." He goes on to say that, "this double chronological principle" entails a fundamental law of evolution, namely, "that man is isolated on the species map, and that no monkey, even a hominid, can transform into a man." We may ask ourselves, then, why he explores just that in *Yzur*, where the narrator attempts to turn an ape into a man by teaching him how to speak—with tragic results.

ESTRADA AND AMEGHINO: CATHOLICISM AND SECULARISM

Upon entering the sciences, Ameghino was subject to the field's rather strict rules, whereby the rise and fall of its members was determined by political favoritism. In the specific case of Ameghino, it is important to keep in mind that at this time Genoese immigrants in Argentina represented a growing threat to the position of the local oligarchy. Unlike Moreno and Holmberg, the brothers Carlos and Florentino were not the children or in-laws of the "old criollo" families, nor did they attend the same schools or move in the same social circles as them. Irina Podgorni has described how the Catholic "Patriotic League," a group which was opposed to the University Reform, saw in Moreno an alternative figure to that of Ameghino. It was on account of this that, for the socialists, Moreno came to be known as an anti-evolutionist associated with reactionary ideals. Since the dawn of the Darwinist controversy in Argentina, Catholicism had come out against the theory of evolution, which did not recognize the Divine Plan in the creation of humanity. In 1880 José Manuel Estrada, in his speech, "Naturalism and Education"—an excerpt of which is included in this section—came out against Darwin, his "villainous consequences," and the "laughable pride" of his followers (among them Ameghino).

The heated controversy between supporters of Ameghino and Moreno rekindled the scientific ferment depicted by Holmberg in his 1874 novel *Two Parties in Conflict*, where conservative and Catholic scientists of Estrada's ilk are referred to as "Estaca" (Stakes), so-called because they are proverbial "sticks in the mud," opposed to the progressive movement of the sciences.

TWO PARTIES IN CONFLICT

(1875)

Eduardo L. Holmberg

CHAPTER IX

In London—Charly And Dick—Dissection of a Monkey in a Room at the Zoological Garden—A Great Discovery, That is to Say a New and Important Physiological Phenomenon—A Letter—"Good Bye"

Let's leave behind for a moment the city of Buenos Aires, the center of this great scientific conflict; let's leave behind for a moment the bosom of its families, where a Rabianist father calls his son *scum* and *riffraff* because he is a Darwinist; let's leave behind the newspapers that are impassioned or neutral to varying degrees, and with that characteristic ease of those who write, let's disconnect from this month that has gone by, let's forget about the Atlantic and the means of crossing it, and let's find ourselves on Regent's Street, one of the main streets, if not *the* main street in all of London, capital of Great Britain.

Today is July 18th, and the noise of the vehicles, the smoke coming from the chimneys, the shouting of the boys selling newspapers, and in general all of the nuisances of a large and populous city, the ancient capital of a powerful kingdom, keep us from staying in the city center for long.

Furthermore it is Saturday—and everyone knows that it is impossible to move about the streets of London on a Saturday, at least until two in the afternoon, which is when all the shops close their doors in order to rest up until the arrival of Sunday, when deep-rooted customs keep the people from having fun. This is why on Saturdays, after two o'clock, everyone is occupied with some sort of *diversion*.

Every half-step we take we run into some money-collecting salesman, those sorts that in Buenos Aires carry the telling name of *Englishmen*, and who are nothing other than personages typically comprised of a man and a more-or-less certain number of bills to collect.

We haven't gone to London in order to do what *London* does whenever it goes anywhere; nor do we have any business with the *Englishmen*, as we not only have no debts with them, but moreover have no desire to incur any.

Let's carry on, then, with our route in spite of the obstacles we may run into, but let's carry on along Regent's Street, towards the south, let's cross Piccadilly Street, let's reach Saint James's Park, from where we will be able to see the King's Palace on the right, and walking around the eastern perimeter of the park we will reach its south-east corner, where, taking a left, we turn down George Street, until we reach Westminster Bridge, from which, if we pause for a moment and look to the south, on the left side of the Thames, we will be able to see, less than a block away, the English Parliament, which is to say the House of Commons first, and a little ways to the south the House of Lords.

Let's pass through the two curves that make up the street at the end of the bridge until we reach the circular plaza that houses the obelisk, some three-quarters of a mile from Westminster Bridge, and heading along down London Row, which runs to the south east, after traveling roughly a mile to the south, we turn down Manor Row and arrive, after having walked more than a league, at the celebrated Zoological Gardens, in one of whose rooms there is a cadaver on a table, and next to this cadaver and this table, a stooped-over man, standing, looking on avidly at he who will be torn apart in just a moment by his able instruments.

This man, of advanced years, wears a white top hat, short dark pants, a light striped vest, a problematically-colored greatcoat, an English nose and a cane as bulky as the hat; but his gaze is that of a great man, and his brow, though at this point it has a slight slope to it, radiates the glimmer of profound science.

He holds in his hand various instruments for dissection. Now he examines the scalpel blades, now the resistance of the forceps, from time to time he sits down only to stand back up again—and then he takes out his watch, has a look, and sees that it is still not time.

He grows impatient—and this is what we most admire, because the unaccustomed person believes that an Englishman must never grow impatient, and all the more so when the appointment in question has been given by one son of Albion to another son of the same mother.

But the one who is waiting there, aside from his nationality, has an additional, unique condition: that of being a great scholar, and, above all, belonging to the Rabianist party, of which he would have had no problem in being named leader, with all of the imaginary responsibilities thereof.

The dissector takes out his watch for the twentieth time, and cannot help but mentally exclaim: "*It wants half a minute*,"[1] but he has not yet finished construing (the words are construed in English) the last word, when he hears calmly and evenly paced footsteps coming towards him down the hall.

At the moment in which all of the clocks in London strike ten in the morning, a human bulk comes between the man with the scalpels and the landscape outside.

—"Good day, Dick," exclaims the man who has just arrived.

—"Hello, Charly Bob, how do you do? How are those lawsuits going, Mr. Magistrate?"

This Mr. Magistrate really is one, from Kent, and having arrived in London the day before, he was unable to resist his friend Dick's invitation to attend the dissection of a monkey in one of the rooms at the Zoological Gardens in London.

It appears that they address one another with reasonable familiarity. All the better; this way they won't waste their time and words conversing about the bones of Labyrinthodonts or Pterodactyls, as many specialists do.

—"So, Master Dick, shall we begin?"

—"I'm ready."

They place the cadaver's neck in a brace, straighten it out well, and get ready to make the incision that some students of medicine in Buenos Aires call a *bib cut*, an incision that begins under the left clavicle and moves down to the lower part of the abdomen, where it follows the curve that runs along the pubis, then turns back upward on the right side, making the same cut on the opposite side of the body.

The two characters—because this is what they truly are—look at one another, fold up the right sleeve of their coats, look at each other once more, and in the manner of two team captains in the tournaments of old, launch forth in a fury.

One of them, Dick, is known to be experienced in the art of making incisions, especially in the bodies of monkeys; the other, Charly, who is less of a knifer, looks on, helps, and meditates.

—"This monkey is a real treasure: it was the last gift that our illustrious friend Dr. Livingstone gave to us," says Dick, as his scalpel reaches the curved part of the incision.

—"What a shame!" exclaims Charly Bob, tearing out his white beard and *smoothing out* his mustache. "If Dr. Livingstone hadn't died, he would have done me a great service."

—"Namely?"

—"He would have sent me an Akka."[2]

—"For what?" asks Dick.

—"To write the final chapter of my theory."

—"Ha, ha, ha! There you are, always worried about your theory . . . ; I examined them in Alexandria. The king of Italy gave me one which may . . . Please be kind enough to hold this forceps here for me."

—"How many days ago did this monkey die?"

—"Three days," answers Dick, while he cuts the fourth rib with a pair of scissors.

—"Three days?"

—"Yes, and we haven't been able to determine the cause of death."

—"Perhaps he wasn't able to get used to the climate."

—"In any case that would only be a cause of susceptibility, to the illness I mean. What we find intriguing is that he shows no signs of putrefaction. Who knows, maybe it was all the cognac they gave him . . . "

—"For what?"

—"To acclimatize him . . . but . . . this scalpel is dull. Are there any sharp ones in that box over there?"

—"Here's one."

—"What is truly amazing is that these veins and these arteries are full of blood."

—"What! It's true! Ergo he is alive."

—"This edge needs to be fastened to the head with this forceps," says Dick.

—"His heart is beating!"

—"Bah! Bah! The animal has been dead for three days now."

—"He may have had a paroxysm, in which case it's possible that the symptoms of this state were confused with those of death itself," says Charly as he passes a sponge moistened with vinegar under the monkey's nose.

But his observation is like the seed of the Anemone: it blows away in the wind. Dick neither looks, nor feels, nor hears. All of his faculties, all of being has been concentrated on one single point.

The movements of his hands, his hands that number among the most skillful of all the nineteenth century's dissectors . . . have ceased to move, as his palm has come to rest on the edge of the table; his eyes have been injected with astonishment, if one can say such a thing. They are staring at the monkey's heart, with that inquisitive avidity of the great scholars.

Dick, —no, Richard Owen, the great naturalist, *Old bones*, as he is called by the English cartoonists—has just made a great discovery.

Charly, —no, Charles Robert Darwin—observes and meditates; and yet, he meditates and observes as only Darwin knows how.

That pongid monkey, with its chest and abdomen open, spilling blood from all the veins and arteries that have been cut, with his heart beating in full view of his two examiners, presents them with a phenomenon that no one prior to them has observed.

For Owen, a Rabianist, this phenomenon is a new and utterly precious bit of information which science will take up for its annals.

For Darwin, it is a magnificent presentiment.

What is it that they have discovered in the monkey's heart?

We have no time to reply, reader, as one of the establishment's workers has just appeared at the door to the room.

—"Mr. Charles Darwin?" he asks.

—"How can I help you?"

—"I've received this letter for you, sir."

Darwin takes the letter and reads hastily:

Buenos Aires, June 20th, 1874
11:00 P.M.
"I have just attended the first session of the Argentine Scientific Conference.
"The Darwinists and Rabianists have argued heatedly.
"The former triumphed, as was to be expected.
"A scholar I was unfamiliar with, Griffritz, administered a near *coup de grâce.*
"The second session is to be held before two-months' time.

HBM Consul
to Charles R. Darwin in England"

—"Owen, goodbye."
"Darwin! Darwin!" exclaims the Rabianist; but Darwin does not hear him—he is running, flying—and a moment later he is lost in the labyrinth of London's streets.

Meanwhile, Owen has been left all alone.

Instinctively he picks up the letter that his friend has left on the dissecting table; but he does not read it. He directs his gaze at the monkey that a moment earlier he tore apart in the name of science.

He steps back, his arms bend, his hands clench, his elbows appear to touch one another behind his back, his hair stands on edge, his eyes and mouth open disproportionately wide, as they are injected with fright and with blood; he wants to scream—impossible!—he wants to move—impossible! The monkey's hemorrhage pours forth in torrents; a moment longer and he will be a cadaver.

Through a supreme act of willpower he moves toward the table and, trying to multiply his number of hands, attempts to tie off the arteries and veins from which spouts the largest portion of blood.

All is in vain. His very speed is an impediment. His clothing, his hands, the table, the floor, everything is covered in blood. An awful struggle; one more second and the *monkey* will be a cadaver. The last drops of blood spurt out. It's all over.

The scholar runs to the door and in the height of the worst anguish possible exclaims:

—"Darwin! Darwin! We've committed a terrible error! Darwin! Darwin! We've committed murder! It was not Livingstone's monkey . . . it was . . . "

But he can't go on.

All of the dramatic sensations produced in less than an hour's time are more than sufficient to visit nervous mayhem upon the scholar's system.

Let's leave him in this state for the time being, and return to Darwin, whom we last saw rushing out of the Zoological Garden, hatless, caneless, and with his hands covered in blood.

In England, as in the rest of the world's cultivated nations, Darwin enjoys a well-earned celebrity, as he is one of the men who has most contributed to the dissemination of the natural sciences in this century, seriously formulating the ideas of Lamarck and other scholars, founders of the great theory known today as Darwinism.

Darwin is English, and he exhibits the speculative character of his countrymen to a remarkable degree, though not in the sense of pounds sterling, but rather in terms of giving a solid, unbreakable foundation to one of the most important questions to stir the scholarly spirit of the age: his theory.

At the age of 17, Mr. Darwin read before the Plinian Society two papers on Natural History, and from that time on Darwin's name began to draw the public's admiration up to the present day, in which it shines as one of the brightest torches of this century's scientific progress.

With these few precedents in mind, the reader will not be surprised that the day after his arrival in London he was invited by his dear friend, the great Richard Owen, to witness the dissection of a *pongid*[3] *monkey* in the Zoological Garden of the capital of the United Kingdom.

Respected by any and all of England's notables, Charles Robert Darwin is in good standing with her gracious sovereign, with all of her ministers past and present, and in all likelihood with the future ones as well.

As soon as he bid farewell to his friend Owen with the laconicism to which we have attested, he ran to the Royal Palace, where at that moment Her Majesty was occupied with matters deeply affecting the interests of her vast domains. Once admitted into the Royal presence, which was not difficult for him to achieve, he expressed in a few brief words the purpose that had brought him there.

—"There, on the edge of the American Continent," he said to Victoria, who had faced him and was listening, "there in the capital of the Argentine Republic, in Buenos Aires, the public spirit has been deeply shaken, awaiting the resolution of the problem being discussed in the bosom of the Scientific Conference. I will not distract Your Majesty with the details, with which, furthermore, I myself am unfamiliar. The Queen of England is not unaware that a theory has been named after her humble subject, and that if, in a city as important as Buenos Aires, his ideas were to take hold, there is no doubt that the moral effect brought about in the rest of the world's nations will have a marked influence on this century's spirit of pragmatism. The advantages that would be reaped from my doctrine's success, particularly by England, are incalculable, because who knows whether one day this success will amount to a glorious precedent for our nation, one that will allow us to wave our victory standard above the main square of that great city?"

—"And so?"

—"My presence at the 2nd Argentine Scientific Conference is indispensable."

—"I see no problem in your departing."

—"Yes, but there is no boat whose speed can compare with that of the *Prince Albert*, with its paddle wheels and double screw propellers, currently anchored at the mouth of the Thames, and as I must arrive before twelve days' time in order to attend that Conference . . . "

—"You hope that I might lend you the *Prince Albert*, if I am not mistaken?"

Darwin bowed.

—"No; I will not lend you the *Prince Albert*, because the *Hound* is even faster."

We do not know whether or not these details are trustworthy, but as they have been passed on to us by an impartial observer, we see no problem in recording them.

A moment later, Darwin was climbing aboard one of the little steamboats that make up the traffic on the Thames, carrying in his coat pocket the Minister of the Navy's order to the Commander of the *Hound* that he obey the orders of Charles Robert Darwin, FRS.

Traveling downstream, he saw a similar steamboat pass by, heading upstream, on whose deck stood a man, who, upon spotting him, shouted:

—"Aren't you Mr. Darwin? I'm an expeditionary sent by Mr. Griffritz, the Darwinist, from Buenos Aires. I've just come from central Africa and I've brought some belongings and reports that Mr. Livingstone, before dying, asked me to deliver to you."

The naturalist heard none of this, since the honking of both steamboats constituted a most noisy interruption, blocking out the expeditionary's voice.

But as he was passing under the central archway of London Bridge, he saw Owen, top hat dented and clothes soaked with blood, who shouted to him:

—"Darwin! Darwin! We've committed murder; it wasn't a monkey, but the Akka given to me by the king of Italy that we dissected! The things that happen when one doesn't wear his spectacles!"

—"Aren't Akkas monkeys?" asked Darwin.

—"No, no, they're men."

—"Well then he'll be one more statistic to add to England's mortality rate."

CHAPTER X

Darwin Aboard the Hound—Crossing the Atlantic—Griffritz's Monologue

In general the English government, as with most enlightened governments, sponsors scientific trips and satisfies the whims of its scholars, a sacrifice that is not so painful in any case, in that all of the expenses eventually make their way back to the assets of the British Treasury.

Are we going to reprimand the English government for this?

We wouldn't dream of it.

What does it matter, in any case, that thousands of subjects perish from hunger and poverty, if in return thousands of pounds sterling can be spent so that a select few might have the pleasure of seeing Venus pass in front of the sun?

What would become of astronomy if opportunities like the 8th of December, 1874 and 1882 were not taken advantage of?

And so no one in England paid any mind to the fact that the pleasure conceded to Mr. Darwin was going to cost a few pounds—since everyone there is keen on scientific advancement, nobody says a thing; and this is why Mr. Darwin continued calmly on his way down the Thames until he reached the river's mouth, where he came upon the *Hound*, whose speed was twenty-five miles an hour. In theory it could reach up to thirty miles an hour with first-rate coal and favorable winds, a truly high speed, though still not so fast it could really be made to clip along were such a circumstance ever to take place.

The *Hound* was a fine steamer that completely satisfied the great naturalist's desires, because, if the speed were maintained at a constant of thirty miles an hour, he could reach Buenos Aires in eight to nine days, which would afford him plenty of time to rest up after the long journey, before the second session of the Argentine Scientific Congress was held.

The Commander of the *Hound* was informed ahead of time and had the boilers heated up, such that when Mr. Darwin arrived the pressure had already risen to fifty pounds and they weighed anchor immediately, everyone took his position, and, opening up valves and cylinders, there began to sound the clanking of the pistons, the screw propellers, the wheels, of the wind whistling in the rigging and bellowing in the sails: sounds all of which served to embolden the giddy hopes held by the naturalist at that moment.

In a word, the *Hound* began its voyage in the most favorable way possible.

A short time later the screw propellers and paddle wheels beat the tepid waters of the English Channel, and the sound grew louder wherever the current ran against them.

—"What a coincidence!" said Darwin to the Commander once he was on board, and the boatswain was hoisting the English flag, and the engineer was opening up the steam boiler. "What a coincidence! Forty years ago, when I was but a young man, I set out for America with Admiral Fitz-Roy aboard the *Beagle*; and now, as an old man, I'm setting out with you for the same place aboard the *Hound*. It seems that I'm being chased by dogs.

—"Or perhaps that you are the one chasing the dogs," observed Commander Clownfield.

—"Dear sir! It's true!" said Darwin smiling with his left Risorius and Depressor septi (as some scholars tend to laugh).

No one heard him say another word for two days. He was thinking up a new theory, in which dogs and monkeys took a more active part than they should have in the genesis of man, and for which he based his ideas on a combination of certain physiognomical observations.

When he awoke from his state of concentration he learned from Commander Clownfield that the *Hound* had made thirty-one miles an hour, its average speed during the two days that Darwin had been meditating in his cabin, such that they had traveled one thousand four hundred and eighty-eight miles, or four hundred and ninety-six leagues. Furthermore, the trade winds, which had blown obliquely through the sails, along with the state of the machinery and the hull, noticeably favored their quick pace.

But let's let Darwin cross the Atlantic, and, as he makes his way to Buenos Aires, let's see what was happening in the meantime in that city.

Since the Darwinists were anxious to clinch their success, and the Rabianists were even more anxious to not let it slip through their fingers, they had made petitions in public and private to hold, at all costs, the second session in order to put an end once and for all to the diverse manner of disturbances that, due to this indecision, were still being incited in the bosom of even the most peaceful and the most sincerely friendly families.

The president stated that by no means would he refuse such a just proposal, and that, if they wished to anticipate the session, holding it on the 28th of August instead of the 30th as had been said at the first session, he could see no problem at all.

Once the proposal had been accepted, the secretary wrote up a brief notice to be published in all of the dailies and bulletins, and on posters all over town. But as he was heading to the presses to publish the notice, he heard a boy selling bulletins shouting: "Shocking murder," "Something," "Something else," "Darwin in Montevideo."

He put the notices in his briefcase and ran to tell the President, to see what was to be done.

And so the notice was changed, and instead of "the 28th," "the 30th" was written, since who could doubt that Darwin's presence in Montevideo indicated that he wished to be present at the second session?

The news spread with that speed with which news spreads amongst the idle, and an hour later, with due alterations, the ill-intentioned were announcing that Darwin didn't want to miss the occasion to see humankind transformed into *monkeykind*, and that he would disembark in less than half an hour.

But all of this was false. It was true that Darwin was in Montevideo, and nothing more. He had made the trip from London to that city in nine days!

Meanwhile, what had become of Griffritz? Nothing could be easier to ascertain. His museum, always open, gives the daring easy access. Sitting before a lit fireplace, one could hear him speak the following

Monologue

"In just three hours the celebrated Mr. Darwin will set foot on the beaches of Argentina, and his name, moving by word of mouth, will find its way to the most secluded reaches of America.

"And what has Darwin come to do in Buenos Aires?

"Am I not here myself?

"Does he think that he is more learned, or more intelligent than I am?

"Am I perchance insufficient to counter not only the weight of the Rabianists' opinion—though in truth the Rabianists are incapable of holding an opinion—but also that of my own proselytes if by some chance they attempted to turn against me?

"Am I not the most learned man on the American continent and perhaps the entire planet? So then why is Darwin coming?

"Oh! Vanity of vanities!

"We scholars are the most intransigent people when we reach the Zenith of our glory during our own lifetimes.

"That is why Darwin must suffer all the weight, all the fury of my criticism, because he has reached the apogee of his fame, and I . . . I have as well. At this very moment my name travels as fast as a meteor up and down the continent and the lands caressed by the eastern sea foams of the Atlantic, and, rattling along the telegraph wire, the name Griffritz spills out across the world like the waters of the Deluge.

"But . . . I forgive him—he's right—my great writings have been archived in this museum, along with my own self, and more than once some curious person must have pointed at me and said, 'Is that man also part of the Zoology collection?'

"If at this very moment Darwin stepped in with the most distinguished of my proselytes and his, I would tell them:

"'Gentlemen:

"'The first Argentine Scientific Conference has been held, the proceedings of which have been written up by my secretary Mr. Kaillitz and corrected by me, and are now perhaps circulating in the great centers of the Americas and Europe.'

"This surely would have some effect on them.

"'In that Conference we'—no: I triumphed; but I would say '*we*' to them because in historical events everyone wants to play a part, even as the losers—'in this Conference we triumphed, as we shall triumph in the second one. That is why you, Mr. Darwin, have come, that is why we ourselves are here, to annihilate the opposition with our presence.

"If we do not win, we are confident in our ability to agitate. If the Rabianists triumph not in discreet debate but in the stupidity of their coreligionists, we can still take comfort in saying: "There are defeats which look like victories," or rather: "There are victories which look like defeats." We still have strength and action in sufficient proportions to stop the winning group from proclaiming the supremacy of its false principles.

"If the Rabianists triumph we shall see the proliferation of the status quo, with all of its excess, with all of its fatuous stability. The sciences will not advance, and even if they do it will be a negative advancement, as I see it.

"If on the other hand we the Darwinists triumph, as is to be expected, since this is the reason why I myself am representing the party from Buenos Aires; if the Darwinists triumph, as I was saying, it goes without question that the entire social norm will have to change, and, either a philosophical revolution of incalculable transcendence will erupt, or an indifference will come about so extreme that we will no longer be able to appreciate the influence of a scientific doctrine on the progress of society.'

"That's what I would tell them, and they would be obliged to give me a round of applause. As for me—though out of tact I would not say this—can I be but firmly convinced that my knowledge contains all aspects of the physical and moral sciences?

"Do I not have superhuman talent, endless good fortune, limitless sagacity?

"Is this not enough for me to be able to help myself to all available resources, moral, intellectual, and pecuniary, in order to attract towards me everything that is susceptible to attraction?

"But I don't want to hear people say that our victory was illicit. Let's leave such means to the Rabianists.

"Estaca himself, that idiot Estaca, couldn't I possibly convert him to Darwinism by offering him *spores* from all of my Polypodys?

"And in the end, what does Estaca amount to, scientifically speaking? One of those elements that arranges other elements around itself, which with all of these strengths working together may come to constitute a certain force.

"Because, although I consider myself to be a great scholar, one must keep in mind that the majority of the public leaves its ability to reason aside, and judges only in accordance with the passion inspired in it by the individuals who make up the parties.

"In that case, what is the point of me dazzling those of my party, if Estaca dazzles a large part of his own?

"And who is Estaca?

"He is the living picture of those great cohorts called 'pretentious ignoramuses,' that plague of individuals who, though on their wings is not written—because they haven't got any—the injury that they cause, as the Arabs tell us, they do bear, in their gaze and on their foreheads, the mark of contagious disgrace.

"Paleolitez . . . no, Paleolitez deserves my full respect, is laudable and praiseworthy . . . but, might one not be able to apply to him that saying of the Great Napoleon: '*Every man is for sale, one need only know his price*'? Everything that Paleolitez aspires to own can be found in my museum and in that case . . . but no . . . Paleolitez in not susceptible to bribery . . . Paleolitez is a worthy rival.

"But . . . what does it matter? Does a grain of sand that passes between us and the sun cause an eclipse? The axis of its shadow's cone is even smaller than its diameter.

"For me it is enough to know that I can interpret nature with the same ease as an Egyptian hieroglyphic or a cuneiform inscription; that I have deciphered all of the *quipus* in my Peruvian collection, along with Galileo's logogriph concerning the rings of Saturn, which no one has been able to read until now, and the discovery of which does nothing to advance modern knowledge; that in seven years I have written fifty volumes that on their own would be enough to characterize a great century, works that later on, when they are published, will force the world's scholars and writers to exclaim: 'From this day on the nineteenth century will no longer be the century of Humboldt, but rather of Griffritz.'"

With this he had said enough.

Mr. Griffritz paused for a moment before a mirror . . . his face was inflamed, and his eyes, in the same state as his face, shot off unusual sparks that refracted in the lenses of his glasses and, brushing against the edges of his hooked nose, met a short distance in front of his face, such that they formed a sharp imaginary dart.

Væ victis! When will he take the floor in the Teatro Colón and so energetically defend his scientific principles!

[. . .]

NOTES

1. TN: In English in the original.
2. OAN: The Akka are a race of men discovered not long ago in Africa. Their particular traits place them so near to the pongid monkeys that we do not hesitate in considering them to be one of the links that should join man and monkey, for which we yield to the opinion given by the traveler Mr. Schweinfurth at Alexandria's Anthropological conference, in spite of the opinions of the renowned Richard Owen, who examined there the ones that had belonged to the lost and daring Mr. Miani.
3. OAN: This word, which appears more than once on these pages, means: [TN: an ape] with a human-like figure, form, or appearance.

CHARLES ROBERT DARWIN (1882)

Eduardo L. Holmberg

SECTION III

Men perish, but man shall endure; lives die,
but the life is not dead . . .

—Swineburne—*Hymn of man*.

Charles Robert Darwin,[1] son of Dr. Robert Waring Darwin and grandson of Dr. Erasmus Darwin, one of his scientific predecessors—and, on his mother's side, grandson of the renowned porcelain manufacturer, Josiah Wedgewood—all of them members of the Royal Society, was born in Shrewsbury on February 12th, 1809. He was educated at Shrewsbury School, where Dr. Butler and later the Bishop of Lichfield were headmasters, and then in 1825 he moved on to the University of Edinburgh, where he spent two years. There he studied marine zoology, and towards the end of 1826 he presented his first two scientific papers, before the Linnaean Society.[2]

He had not yet turned eighteen.

Later he studied in Cambridge, at Christ's College, where he received his Bachelor of the Arts degree in the Sciences in 1831. In the fall of that same year, when preparations had already been made for Fitz-Roy's expedition around the world aboard the *Beagle*, the captain offered, for lack of an extra cabin, a place in his own, to any naturalist who wished to accompany him.

Darwin offered his services without any monetary compensation, but with the condition that he be free to have at his disposal the collections he would gather on the trip.

On December 27th of that same year, the *Beagle* set sail from Davenport. The main objective of the journey was to study the hydrography of the southernmost regions of South America, and, given the nature of the work that Captain Fitz-Roy would have to undertake, it is easy to understand how the young naturalist, just twenty-two years old, had the chance to carry out his important studies in the regions they visited.

Once they arrive in Patagones, Darwin heads on horseback to Río Colorado, where at the time Rosas had his general headquarters. He receives him well, provides him with passports, and lets him make use of the outposts that were established for the purpose of keeping the military expeditionaries in contact with the government. Darwin reaches Bahía Blanca, crosses over and studies the Sierra de Currumalal, goes to Tapalquén and from there to Guardia del Monte, wherefrom he heads for Buenos Aires, where he stays at the home of the well-known Lumb family. This took place in 1832.

I need not remind you that his long and painstaking journey on horseback was not sterile for the sciences. Darwin was one of the first people to study this Province of Buenos Aires.

We will skip over his travels along the Paraná, through Banda Oriental, and his later stay and studies in Patagonia, the Falklands, and in Tierra del Fuego, where he made many very interesting observations. Let us open up his travel diary and point out a coincidence.

"*April 20th, 1834.* We passed the islands and set to work. Our regular day's march, although it was hard enough, carried us on an average only ten miles in a straight line, and perhaps fifteen or twenty altogether. Beyond the place where we slept last night, the country is completely *terra incognita*, for it was there that Captain Stokes turned back. We saw in the distance a great smoke, and found the skeleton of a horse, so we knew that Indians were in the neighbourhood."

This note corresponds to the exploration of the Santa Cruz River, suspended on May 4th of that same year.

Exactly 48 years after it was written, the brain that had dictated these words ceased to stir; meanwhile, the study of the *terra incognita* was to become one of the finest laurels worn by one of Argentina's most courageous explorers: Francisco P. Moreno.

The travelers head for the Pacific, and over the course of these explorations our naturalist, always collecting and observing, develops the seed of the doctrine that would go on to immortalize him. One of his most meticulous studies is that of the Mendoza Range, given that it deals with our country. It is a study for which Alfredo Stelzner was unable to find sufficiently laudatory words of praise, given the level of sagacity unleashed by Darwin in his analysis. Later on, in Polynesia, his observations as to the formation of coral were to further enlarge the pedestal he was carving out for himself.

After having been gone for five years of tireless work and fertile labors, the explorers arrive in England on October 22nd, 1836.

Early in 1839 Darwin married his cousin Emma Wedgwood, with whom he had several children who, cited on more than one occasion in the distinguished scholar's papers, and whom he has counted among his best helpers and collaborators, will surely live up to the great name that has been passed down to them by three generations.

Beginning in 1842, Darwin, whose fortune sheltered him from all work other than his studies, lived in Down, Beckenham, in the county of Kent, where he was named magistrate in 1874.

His first publication was the diary of his travels, one of the most interesting works that can be read in this genre. This book, corrected and expanded in 1845, was then stereotyped and in spite of its many editions it has not been modified further.

Darwin had sent the great collections gathered over the course of his explorations back to England from time to time, to his friend and teacher Professor Henslow, who had been able to nurture Darwin's hereditary seed with his lessons. These collections were distributed amongst various scientific institutions, which motivated a series of foundational publications whose content should be the ABCs of all Argentine naturalists. The fossilized mammals were published by Owen, the present-day ones by Waterhouse, the birds by Gould, the fish by Reverend L. Jenyns, and the reptiles by Bell. Waterhouse, Walker, Newman, and White worked on the insects, and Hooker (Dr. J.), in his study of Antarctic flora, described the plants, an addendum to which he also produced. Henslow published those collected in the Keeling Islands, and Reverend Berkeley, the cryptogams.

Darwin himself took part in editing the specialized works that make up the large series to which we have alluded, carried out with the liberal aid of the Lords Commissioners of the Treasury, who contributed to the high costs of publication with £1,000.

It would take up far too much of your attention to list here of all of Darwin's specialized and general works.[3]

Suffice it to recall from among them:

The Voyage of the Beagle.

On the Origin of Species.

The Variations of Animals and Plants under Domestication.

The Descent of Man and Selection in Relation to Sex.

The Expression of the Emotions in Man and Animals.

On the same day that Buenos Aires received the news of his death, the French translation of his last book, *The Movement and Habits of Climbing*

Plants, arrived as well, and judging from its title and the summaries of it, since we have not yet had enough time to read it, it promises a world of new findings for plant physiology.

In order to expound, even in general terms, the doctrine that today carries the name *Darwinism*, we would need much more time than we have today. However, since we must at least mention it, and offer a sketch of it, we will allow ourselves to cite a few examples that illustrate it. We will take these examples from our immediate surroundings, from the natural world of Argentina, from our gardens, from our birds, from our struggles and, if you wish, lovely daughters of this land, from your very faces.

In his foundational book titled *On the Origin of Species by Means of Natural Selection: Or, the Preservation of Favoured Races in the Struggle for Life*, Darwin lays out his material in the following manner:

Chapter I. Variation Under Domestication
Chapter II. Variation Under Nature
Chapters III & IV. Struggle for Existence[4]
Chapter V. Laws of Variation
Chapters VII to IX. Difficulties on Theory[5]
Chapter X. On the Geological Succession of Organic Beings
Chapters XI & XII. Geographical Distribution
Chapter XIII. Organic beings' classification or mutual affinities in their adult and embryonic states[6]
Chapter XIV (last). Recapitulation and Conclusion

We cannot discuss every one of these chapters, because some of them deal with *delicate* subject matter, particularly for feminine ears, which are not used to scientific candor, and because others would require an extremely long and methodical explanation. [. . .]

Thus we will take up, as we said earlier, a few well-known examples.

Variation Under Domestication

Due to causes which we take to be unknown, though research into them is perfectly within the domain of science, several leaves on one of our rose bushes, on one particular branch, exhibit yellow spots as they develop from their bud. Gardeners call these variegated leaves. This is a common phenomenon, which we can observe regularly in camellias, violets, Cape honeysuckle, and other common plants.

What a shame not to be able to grow a rose bush with variegated leaves! But we do not know what to do, and leave the branch where it is. Still, we are able to think up with a possible procedure. We cut the branch and plant it in the ground. After a certain amount of time it sprouts roots (roses start easily from cut branches) and its buds open up and form new branches. To our great surprise, only one of them has yellow-spotted leaves.

But how might we obtain an entire rose bush whose leaves are all spotted?

Cumulative selection. This one branch, which we can conventionally call "second generation," is separated and planted, a procedure which we continue until the entire rose bush has spotted leaves.

Most of the plants in our garden that have variegated leaves were formed in this way under cultivation.

What we have done here is not marvelous. It is not abject. It is not base.

What is it then?

It is artificial selection. This is the *first chapter* of Darwin's book, about the origin of species.

Let us move on to animals.

Among the many canaries in a birdcage, some sing poorly, and others sing well.

Let's separate them, ridding ourselves of the former, and keeping for ourselves only the latter, so that their offspring, by inheritance, will have this paternal quality, namely that of singing well. Those that do not sing well will be rejected until, in the end, we have a race of excellent singers. In this case, their song is the preferred trait, independent of their plumage. If plumage were in fact the trait we wished to modify, attempting to obtain, for example, a white pigeon with a black tail, we would avoid getting the one that is most like the type we have in mind mixed up with the others, and, accumulating approximations to it, we would arrive at said type. Through this procedure such extraordinary results are obtained that we would find them unbelievable, if we did not have the opportunity to see them every day.

It was in these studies of artificial selection, to which Darwin dedicated special attention, that he discovered part of the basis for his explanations, gathering countless data to this effect.

Variation Under Nature

When man exercises his faculties in the modification of domestic plants or animals, he is producing artificial selection; but man himself is prone to variation, and is capable of carrying out on his own species that cumulative selection that we spoke of when discussing the variegated rose.

Is there any well-known example of natural selection in mankind?

Of course! The Spartans and, to a lesser degree, the Indians of North America.

What did the Spartans do? Any child who was born without a perfect form was tossed to the wolves of Mount Taygetos.

What was the result of this selection? The creation of a vigorous, healthy, beautiful race, strengthened by exercise, but highly brutal.

Among us this sort of selection does not exist.

If a child is born with defects, what can be done! It is unfortunate, but we do what we must so that the child will live and he does. He has his congenital defects, but can amount to something regardless, as was demonstrated by the poet Tyrtaeus, who, despite his bodily defects, sang so well that the Spartans were emboldened by his songs and won a battle on which they had given up all hope. This child's descendents will most likely have defects as well, and, as Haeckel observes, all of these defective men, due to their extreme weakness or any other cause, are no good as warriors. The good ones are big, robust, with muscles like a cannon. These are the ones that can and must die, these are the ones that are heroes, like Frederick of Prussia's men. These are the ones who must stand before the gunfire, while the rest serve only to perpetuate their kind!

We have not undergone military selection, as Haeckel terms it, to the extent that Germany has; but we might still find some trace of it in certain data that we have not yet been able to obtain, but which become available in the relatively near future.

It is almost certain that if one were to compare the average height of third- to fifth-generation men from Buenos Aires, meaning the province's fathers and grandfathers, it would be lower than in the other provinces of the Republic.

And this is understandable. War is our favorite pastime!

Since it seems that all around the world war is used for sport, humanity degenerates. The men with the worst traits are useless as solders, but good enough to reproduce, and will carry on this degeneration, and diseases will spread. But this is unimportant.

Let us put forth an example, which, though somewhat . . . Spartan, can still be taken as an example.

Due to some flight of fancy, men come to prefer blue as the color of women's eyes, while brown or black eyes are mercilessly rejected.

After several generations all women would have blue eyes.

If blonde becomes women's color of preference for men's moustaches, all men who wear moustaches will have blonde ones.

Darwin did not dare venture so far in *On the Origin of Species*, but as we have seen he implied it later on, with *The Descent of Man*.

In the delightful pages that Mantegazza wrote about the women of Buenos Aires, and that could be said to refer to Argentine women in general, there is grounds for an interesting study into the principle of sexual selection. [. . .]

Since there is a higher foreign population in this province, one which is higher still within the capital itself, the women of Buenos Aires undergo visible modifications with each passing day, such that there will come a time when the women of Buenos Aires will be so lovely and graceful that another Mantegazza will have to come along and sing a new ode to their beauty. Here there is no prototype. There is just a certain indefinable beauty about them. In some Argentine provinces Hispano-Arabian features have been preserved with a certain ethnic purity that allows one to trace the traits of their inhabitants back to one Spanish prototype or another. This is doubtlessly going to disappear, as is already happening in Buenos Aires and also, to a lesser extent, in Rosario.

In some European countries dowries are used, and since there a dowry is, for a woman, a ninety-nine percent guarantee of finding a husband, the latter spends his time staring at the dowry, which poses a lively obstacle for the selection of personal traits: money.

Here dowries do not exist as a civil institution, though it is surely not forbidden, nor must a woman with a fortune necessarily be lacking in great personal advantages.

Yet, since in general a powerful agent comes into play, namely the balance of wills, reciprocity in love, in relation to the selection of beauty, intelligence, and grace, the result of this means of selection is a beautiful, intelligent, and graceful template.

The disruptive element, ugly men like us . . .

But gentlemen, this matter is growing much too complicated.

The theory has been laid out.

It can be resolved.

This is, in essence, the *second chapter* of Darwin's book.

We will now give one last example to finish up.

Let us have a look at the fundamental question: *The struggle for life*, as Darwin terms it. The issue is a plant that has invaded the Province of Buenos Aires, and the study of which is of great importance, whatever the results of this research might be.

Allow me to cite a work in progress (*Trip to Tandil*, by the author of these pages):

"Among the varieties of flora in Tandil,[7] there is a unique species which stands out among them. It is a *thistle* that has begun to spread across the province in the last four or five years, coming, it would appear, from the western regions and which, given its rapid diffusion, its characteristics, its robust growth, and the facts of its structure, threatens to take over the position of the physically similar "Cardoon" (*Cynara cardunculus*) in our fields, and of the "Milk Thistle" (*Sylibum marianum*) which predominates in that region. Three years ago I observed it for the first time some two leagues north of Navarro, and several inhabitants of the region, to whom I expressed my surprise at having found a thistle that I had never seen before, told me that it had produced the same effect on them, as they had not previously observed it there. A short time later, when I was with Moreno at the Museum of Anthropology, Mr. Don Pedro Pico, who happened to be present there, expressed the same surprise, telling us that he, who had traveled through this province so many times in all directions, had never seen a certain type of thistle that he had recently spotted on the western side, and the way he described it to us coincided with the species I have referred to. Lately I have found it in rather extensive clusters near Tandil, and in various points throughout the region lying between this town and Buenos Aires, where it is already present within the capital's city limits.

"Neither Grisebach in his *Plantae Lorentzianae* or *Symbolae*, nor Lorentz in his *Flora of Entre Ríos* or *Expedition of General Roca to the Río Negro*, nor Echegaray in his *Florula of San Juan*, nor Hieronymus in his various papers, nor Berg in his *Exotic Plants*, have pointed out the presence of this plant in the Argentine Republic up till now.

"From where does it come? Is it a hybrid plant? Is it a spontaneous product of our fields? We shall dismiss the first idea because the plant does not reproduce the traits of its likely progenitors, and the second because it is simply absurd.

"Its highly branching stem produces an innumerably higher number of flowers than the other thistles. This means that it produces an immense number of seeds, and its propagation is better assured than the others. Its dark-green colored leaves, lightly covered with a whitish down, especially on the underside, and with abundant and extremely sharp spines, ought necessarily to keep animals from eating them. And it must be so, for I have never seen a single one set upon by them.

"Given the conditions or characteristics of this thistle, its startling multiplication, the means for its propagation that exist on our soil, the fact that it excludes other vegetation wherever it grows, the established struggle between this and the other thistles, which during a part of the year constitute our cattle's only sustenance, I believe that I am authorized to call the governments' attention to this plant, which is linked to the prosperity of our nation.

"What I have here is a brilliant study for a Darwinist.

"After writing the foregoing lines I made their content known to Francisco P. Moreno, who, a few days later, came to see me in order to tell me that he had taken up the matter, because it really was of vital importance for the country. He had consulted a wealthy rancher from the province, who told him that the animals did eat it, and that if this was true my fears were unfounded. What very well may have happened is that it was a plant introduced with the Lincoln sheep, which were recently brought here, and which were driven through the Navarro side. This coincidence caught my attention and after having a look through the flowers of Scotland and England, it seemed to me that the plant (as I remembered it) had traits similar to those of Linnaeus's *Cardus acanthoides*.

"A few days later I notified Mr. Pico of these findings and he held that Moreno had been misinformed, that on the ranches the thistles were not being eaten and that my observation had been correct. He assured me that one of the country's leading ranchers had ordered all traces of the plant to be eradicated on his land.

"To the name *Black Thistle*, which according to Moreno is used in the countryside, Mr. Pico added the name *Devil's Thistle*, by which it is also known."

Here ends the manuscript to which we alluded.

For someone who is used to seeing things for what they are, and how they are, independently of their relationship to the world, this plant is called either *Black Thistle* or *Devil's Thistle*, and animals either eat it or do not eat it. This is positivism stripped down to its essence. But the doctrine with which we are concerned would be very poor indeed if it did not teach its initiates to study at length the absolute concatenation of all things and of all causes.

Let us formulate a proposition, supposing that the plant in question *harms* our animals and that they, for this very reason, do no eat it.

What relationship exists between the Black Thistle or the Devil's Thistle and the prosperity of the Argentine Nation?

1st. The *Black Thistle* is a plant which is surprisingly well-equipped to drive out other types of thistle.

2nd. In practice its upper hand will drive out the other types of thistle all across the Argentine Republic.

3rd. Since our cattle live primarily off of those thistles during part of the year, and since they do not eat the *Black Thistle*, they will have nothing to eat and will starve to death.

4th. Any loss of life among our livestock causes a rise in the price of consumer products and, as the Republic's almost sole source of wealth is its cattle, the country will go hungry and will be unable to prosper due to the death of its livestock brought about by the *Devil's Thistle*.

We could carry on with this chain of events, beginning by blaming all of those who spoke of the moral and material progress of the Republic, whose writings incited the importation of the Lincoln sheep with which the thistle was most likely brought over.

Let us suppose, on the other hand, that the thistle is excellent, which unfortunately is not our opinion, and that the animals eat it, and, moreover, that it is better than the Cardoon and the Milk Thistle.

In this case, we can repeat the first and second propositions.

1st. The *Black Thistle* is a plant which is surprisingly well-equipped to drive out other types of thistle;

2nd. In practice its upper hand will drive out the other types of thistle all across the Argentine Republic;

3rd. Since the *Black Thistle* exhibits, in terms of our agriculture, greater advantages than the Cardoon and the Milk Thistle, its propagation, driven by its superior qualities, is bound to be extremely advantageous for livestock;

4th. As the majority of our country's wealth is constituted by its livestock, it is evident that by improving their conditions the country is bound to progress.

We could, supposing this favorable scenario were the case, carry on with this chain of events, beginning by enthusiastically congratulating all of those who spoke of the moral and material progress of the Republic, whose writings incited the importation of the Lincoln sheep with which the thistle was most likely brought over.

This is, then, a case of a *struggle for existence*, the detailed study of which is, to our mind, of great practical importance, as all of the people affected by the solution to this problem, in either point of view, will have understood.[8]

What does *Struggle for Existence* mean, then?

Darwin here applies the *Law of Malthus*.

"Population increases in a geometrical ratio. Subsistence increases only in an arithmetical ratio."

This doctrine contains both the seed and the lifeblood of all manner of attacks, directed at those economists who do not observe Nature at work, and especially at those philosophers who presume that in all things they will find the mark of some act of Providence.

And so, since the *Law of Malthus* is one of the principal arguments against the idea of Providence, and since this idea is too deeply rooted for us to attempt to dislodge it with this humble exposition, we must show that the *Struggle for Existence* is one of the principle arguments in favor of said *Law of Malthus*.

Let us suppose that one *Black Thistle* plant produces five thousand seeds, which then fall all around it. Will all of those five thousand seeds develop? Surely they will not, because a large number of them will be eaten up by the birds: doves, ducks, partridges, etc.

Let us allow for two thousand to survive, for the dust to cover them, and, when the time comes, for them to germinate and grow. Yet can they all grow in the same small area? No. Only four may grow, which is to say that one thousand eight hundred and ninety-six little plants must die. And which ones will survive?

This is where *natural selection* comes along to explain which of them, in this *struggle for life*, battle for space, fight for nourishment, are the ones that will survive.

Out of the two thousand plants, many, for instance, were born weaker than others; these ones die first, because the others, which were stronger, defeated them in the struggle.

In short, let us accept that the last four were stronger than the rest.

Being stronger does not mean having the greatest advantage; nor does being weaker mean having the worst.

As a general rule, a weak, sickly, delicate child is the object of greater care on the part of his parents, whereas the strong, robust child does not receive this sort of attention. He catches pneumonia and dies, while the other one, forever swaddled and dragging out his miserable life, reaches old age.

In a plan conceived by Nature, with true intelligence, with equity, the strong child would not have had to die. This inversion of the natural order, when it does away with selection, is what is called Providence.

Returning to our plants, we now have four that have survived. Their advantages, whatever their nature, have an organic expression: a longer leaf, a more robust root, more compact tissue, etc. By inheritance these traits that have produced an advantage will be passed on, though not in all individuals; but as long as some individuals have them, they will be the victors whenever a struggle arises.

This unceasing acquisition is accumulated over centuries, thousands of centuries, and since the struggle is constantly being renewed, there are more and more privileged, chosen individuals, and the rest die.

This is one of those truths that imposes itself so categorically upon our understanding that we cannot free ourselves of its influence once we have considered its meaning and its importance.

This slow, gradual, imperceptible modification, like the drop of water that over time drills through hard rock, whose powerful action we cannot recognize in just one day or two, but which, with the passage of time, is made clear to our senses, is so evident that only caprice, lack of observation, or the systematic attempt to deny it, can fail to discern it. In such cases let us not attempt to argue against it, as we would betray a veiled intent, which will not necessarily find a social environment of sufficient refinement to suppress a rebuke merited by all those who seek to prove themselves correct through dialectic and not through facts, which are truth itself.

Let's say that in each generation one or several organs are transformed. Though from one generation to the next this transformation may be imperceptible, for our purposes even the most elemental observations are sufficient, provided they have revealed to us these transformations along with the causes that brought them about. We can then accept without hesitating the existence of a force, of an impulse toward an indefinite variation that, over time, will completely separate, within this type, the descendents from their progenitors. If the intermediates disappear, then these progenitors will come to occupy the place reserved for *dead* or *extinct* species.

This is how species *die out*.

However, the life that was their heartbeat has been passed on to their descendants, like the luminous trail left behind in a measurable line by an aerolite. Would we say that they are two trails when a tower disrupts, in our view, the line's continuity? . . .

In the animal kingdom we have the Indian.

Is the Indian's cause just?

Arguing without too much dialectic, the Indian defends his land, which we have stolen from him, he wounds us, he kills us, he robs us.

Does he do right? It is clear, or not. He fights for his life.[9] But since the laws of nature work most noticeably, in their greater manifestations, upon larger groups of humans, rather than on individuals, all the opinions of Providence, justice, equity, brotherhood, which are nothing more than opinions separately rooted in each of us, fall apart in the presence of that common manifestation which is, in a way, the law of nature; —and we Whites, the civilized ones, the Christian ones, the ones armed with Remingtons, do away with the Indians, because the *Law of Malthus* stands above all of these individual opinions, which may be truly excellent, but which, either because humanity is still far from being civilized, or on account of any other cause, don't flesh out—and so, as we too fight for our lives, with good ideas, with good weapons, with good resources, we do nothing more than put into practice our upper hand.

—"Do we do right?" This is a question.
—"We fight for our lives." This is an answer.

Reason, lastly, is a victory of organic progress. But victory, in any form, is a *reason* that overcomes all progresses.

Crevaux in Pilcomayo. His death is an immense, irreparable loss. It is the struggle for life!

All of this, gentlemen, is Darwinism.

This exposition is incomplete, it is true. Others, with more science, could have presented it more brilliantly, but not more honestly.

There are many spaces to fill, many obstacles to overcome, many worries to do away with.

All of these will be filled, overcome, and done away with.

It is the *Law of Malthus* applied to Feeling and Reason, which has given us Darwinism, a great doctrine, whose principles, adapted to all forms of human activity, will surely bring us to perfection, that perfection we have dreamed of and that we should have attained by now, were it not blurred by progress, and because we have not rid ouselves of many of our simian ancestors' traits.

And at any rate, have our feelings made us any better?

We have been Jews for 4,000 years; Christians for 2,000 years; Mohammedans for 1,000 years; and yet . . .

As rogues, are we any less or more so as Jews, Christians, or Mohammedans?

As fools, do Mohammedans, Christians, and Jews differ so much?

As fanatics, is there much difference between Christians, Jews, and Mohammedans?

Do we find less cruelty, less barbarism, less stupidity in the wars of Jews, Mohammedans, and Christians?

Do lowliness, baseness, perfidy, and all the worst spines that bristle deep inside the *Steward of Creation*, do more good in some than in others?

Do the virtues that adorn us shine brighter now than they did in India four or five thousand years ago? Is the average Christian more hospitable than the Mohammedan or the fetishist Araucanian? . . .

We are deluded if we do not feel the throbbing of the races that came before us in the oscillation of time! We still believe that *just and good* belong to the realm of the absolute.

Feeling exists, this is a fact. But its nature as part of the mind, subordinate to each person's organic emotivity, is as variable as individuals themselves and as the circumstances that surround them, and therefore it cannot be imposed as the basis of social organization, much less as a source of judgment.

It varies along with its context, it transforms along with common sense, and its greatest eruptions fade away in relation to the interests that assail it. And this inconstant faculty is the one that governs human societies?

Let's educate it. Let's give it a religion, give it something to love, give it anything that will keep it from growing idle; but let's teach it to bow down before Reason.

Reason, as expressed in science, is the only trait that distinguishes men from the animals.[10]

Monkeys make all the faces we make.

Parrots speak all the languages we speak.

Dogs, horses, pigeons, and cats share all of our feelings, all of our faculties, and fight for their lives just like us.

These are the discoveries of Darwin, these are the proofs used by that Darwin, one of the greatest figures of the nineteenth century, obliged, so that the world could read his book, to present it with the *approval* of a bishop, who had no idea what he was holding in his hands.

Yes, gentlemen, and alongside Reason let us leave a place for Will, that brave and subordinate sister, without whose powerful strength I would not have been able to present this paper, carried out in the midst of the struggle for life.

Here's to Darwin!

NOTES

1. OAN: All of the information pertaining to Darwin's family and to his first years of life are from a paper published in Issue 240, Volume X, p. 79 (June 4th, 1874), of the English periodical *Nature*, titled, "Charles Robert Darwin" and attributed to Asa Gray. Concerning his stay in America I have followed his *Voyage of the Beagle* (Barbier's French edition, 1875).

2. TN: Holmberg has confused the Linnaean Society, the premier society for natural history and taxonomy in Britain, where Darwin first presented *On the Origin of Species*, with the Plinian Society, a student natural history group at the University of Edinburgh.

3. OAN: I do not find it trivial to repeat the list of Darwin's works here, just as A. Gray has given it to us in the cited paper from *Nature*, p. 79. [EN: I have only reproduced the General Works]

Journal of Researches into the Natural History and Geology of the countries visited by H. M. S. Beagle. 1845.

On the Origin of Species by means of Natural Selection. 1859.

This was preceded by a sketch, entitled "On the variation of organic beings in a state of nature"; published in the *Journal of the Linnaean Society.* Vol. III (Zool.), 1859, p. 46.

The Variation of Plants and Animals under Domestication. 2 vols. 1868.

The Descent of Man, and Selection in Relation to Sex. 2 vols. 1871.

The Expression of the Emotions in Man and Animals. 1872.

[. . .]

4. TN: Chapter IV in the original is titled, "Natural Selection."

5. TN: Chapters VII, VIII, and IX are titled, "Instinct," "Hybridism," and "On the Imperfection of the Geological Record," respectively.

6. TN: Original title of this chapter is: "Mutual Affinities of Organic Beings: Morphology: Embryology: Rudimentary Organs."

7. TN: Tandil and Navarro are both cities and *partidos* (essentially counties) within the province of Buenos Aires.

8. OAN: I am indebted to Dr. C. Spegazzini for the technical name of the *Black* or *Devil's Thistle*, which is: *Cirsium lanceolatum.* [. . .]

9. OAN: What a sad obsession, to sink each and every question into the darkness of Ethics. Pedro hits Diego with a club. Does he do right? I do not know. What he *does* is hit him with a club. There is no room for doubt on this point. At the trial, five hundred will think that he *does right*, and five hundred that he *does wrong*, proof that humanity will always have the grounds for an argument. It is *good*, because this way humanity won't live out its days in silent boredom. Pepper isn't necessary, but it is pleasant for those who have a taste for it.

10. OAN: With necessary restrictions. Anyone with any reasonable background in matters of anthropology will understand that I am only referring to the superior races of men and not the lowly ones, by which I mean those that, like the Hottentots, find themselves in the lowest intellectual level. I will insist on this point. To conclude I will repeat what I have said elsewhere: "Not all men are equal."

THE STUDY OF SOUTH AMERICAN MAN

(1878)

Francisco P. Moreno

I

No study more worthy of the human spirit than that of man himself, and yet, it is the one that has least preoccupied him.

Only now—since he has come to dominate all the inferior organisms through the conquests of his intelligence, and since he has become convinced of his preponderance over them as the most magnificent of nature's creations—has man abandoned the idea of his perfect and semi-divine origin, and paying heed to what science teaches him, has stepped down to occupy his proper place on the animal scale.

By discarding a mistake that had exalted him, he has established his true greatness: though he has an animal origin, he has created by his own strength—without recourse to external forces, in a life of unceasing struggle, from the outset divested of all the qualities with which the other organisms are equipped for their preservation and defense, through cataclysms that destroyed a large part of those creatures, and undergoing innumerable social metamorphoses—civilization itself, the magnificent offspring of his genius, which takes shape through him and reveals to him all of the creative forces out of an absolute chaos.

Since primitive times, a sublime harmony has guided and led the way for the succession of all of earth's organisms throughout all of the geological epochs. The law of perfectionism that governs them, and that paleontology has revealed, shows us that living things did not suddenly appear on the earth, isolated from one another, but rather are intimately connected with those that came before them in the theater of life.

Nature's magnificent plan for the formation of the organic world appears at first glance to be unspeakably complex, in terms of the forms of the creatures that have resulted from it. And yet if we have another look at everything that science has demonstrated, we can bring them together in a single unit. All of the animal transformations, starting from the most rudimentary and nearly imperceptible organism, which is barely distinguishable from inert matter, over an incalculable period of evolution, and guided along by that plan and impelled by the creative forces, have led to the development of the human animal, who brings together those elements and who is able to contain, within his physical makeup, the living organization of everything that has appeared on the face of the earth.

In him Nature has brought together all of its creations that have appeared and disappeared, and to his animal form has added something more, emanating from its very own perfection, that allows him to go so far as to understand Nature so as to admire it, even though it is not for him to inquire as to the causes of its origin.

If man's faculties have succeeded, thus completing their own development, in tracing out the succession of animal forms, then it is no small conquest for the spirit, which must follow this wide path, and not attempt to rise up through intellectual speculation to that which will never be deciphered for him.

Let it suffice for us to familiarize ourselves with the gradations among the living organisms, the genealogies of the creatures that preceded our appearance on the stage, and discover the differences or analogies that exist between them. Let it suffice for us to know that from the simple monera, life has been successively perfected, be it through evolution or through separate creations, and has resulted in man, and that he, from a humble ovum, acquires the complications that bring about his development and his being. But let us not attempt to find out what force it was that created the monera and the ovum.

Let us be content to gather, for the purpose of our study, new facts, and we will have carried out our mission. Another generation will weigh them and find new laws that better govern the transformation or change of primitive forms.

As for me, I respect all natural systems, all of the theories that distance themselves from laws that are subject to animal nature: some of them the daughters of observation, others of pure fantasy; each of them reflections of man and as such subject to a never-ending variability.

I will not endeavor to investigate whether or not the laws that govern the genesis of man, or the geneses of any of the genera of creation, can be applied to the totality of his works, recasting them as one original unit, and as the products of branching and of selective refinement over time. I will not investigate whether or not the origin of the current animals is the product of immediate descent from other more inferior animals; or whether they are the fruit of separate creations that were progressively perfected.

I am only going to take up man, as the most perfect being on the organic scale, without looking into his precursors, and study him broadly in physical and moral terms, from the time of his appearance on the earth up to the present moment, an era that is rather short, I might add, in the sum of all of life's ages.

Linnaeus's *Homo sapiens* is clearly the most perfect animal on the zoological scale, when judged according to the admirable law of gradual attainment of perfection. The very type of succession that we find in the creation of the animal world can be seen in man's own intellectual creations.

The stages that the human spirit has passed through in order to obtain his current level of perfection are the same as those that an inferior living thing must pass through in order to move up to a higher level of advancement.

We see the first glimmer of the conscious application of intellectual power in the use of fire, or in the rough carving of the first stone knife or axe, or in the even more remote staff or club—this can be compared to a common monera. And today's human activity, which strives to fight and overcome the elements, can be compared to the best-equipped creature before the appearance of man.

Just a few short years ago there was still an almost total ignorance as to how humanity had progressively developed from the time of its first appearance on the earth, and as to what phenomena had taken part in its refinement.

Several sciences studied man's physical constitution and the morbid elements that afflict him; but in isolation, without looking for affinities of race or of species. Racial classifications were more hypothetical than real, and observations about the customs, practices, and languages of a people were taken more as a grounds for curiosity than as real ethnological study. Their history held the key to the ethnography of a nation.

Anthropology, the science which was to bring this ethnography to light, took its place at the head of all the scholarly disciplines springing from human intelligence. It entered with full force, occupying the highest level from the outset, sure of its undeniable importance. It overcame the indifference of those who were slow to catch on, and gathered a momentum that came as a surprise to many, but which had been prophesized by those who foresaw its rapid march to the forefront of human knowledge. It took advantage of this knowledge wherever it pertained to natural history, as a science of observation; and of all the social advancements to whose study the elevated questions of philosophy are tied.

The emergence of anthropology was inevitable. The human brain, of whose substance and activity thought is a property, needed, as part of its evolution, to widen its sphere of action in accordance with the law of progress. Thus, the study of man by man completed the picture of the branches into which knowledge is divided.

The desire for progress finally led man to inquire into things he had not yet been given the chance to know, spurred in part by the previous century's great scientific expeditions into unfamiliar regions, which revealed new men, ancient civilizations, and unknown languages that began to be appreciated by the period's state of consciousness. They gave notice to the world of the mysterious ruins of Mexico and Peru, those of Egypt, Assyria, Persia, and India, buried under the sands that hid them from the populace's indifferent gaze, but which the keen eye of science observed. Others had been covered by blankets of plants, which, by stealing them away from the profane light of day, preserved their erstwhile appearance. Gigantic and magnificent ancient trees held tight within their branches, in the middle of the jungle, the statues, temples, palaces, and tombs of past generations extinguished by time. Splendid vines dug their roots into the cyclopean inscriptions, in a state of confusion lesser than that of the first man who admired them after they had been lost to history for thousands of years, and who tried to interpret them.

The science of human progress at that point began in earnest; the serious investigation of the past began, and each day new links were added to the great chain that tells of its progress. Over time prodigious eras of civilization were revealed, in places that had up till then only been considered lands of primitive nomads. Almost unnoticeable at the outset, the progress of this science, once it had discovered such elements as these, gained considerable momentum. The presumptions held, by those who had studied history's antiquity, about a hereafter for humanity, gave way to facts, and to phalanxes of brave researchers, as wise as they were patient, who took on the task of straightening them out.

Science scrutinized the past. Del Río, among others, later commented upon by Cabrera and Dupaix, explored and studied the magnificent buildings at Palenque. Humboldt and Waldeck showed, the former with his brilliant descriptions, the latter with his vibrant watercolors, both of which are found in the great New World album, the ruins hidden away by the jealous lands of Montezuma and of the Incas.

The scientific commissions in Egypt produced monumental works, rivaling in grandeur the creations of the ante-historical[1] dynasties up to that point, with which the genius of Champollion, Lepsius, and so many others pushed history's infancy back thousands of years into the past. The stones spoke and told the wise men of today the things that the wise men of five thousand years prior had done, when in what is now civilized Europe men were savages, just like the ones on our pampas.

In Asia, that same desire uncovered Assyrian and Persian ruins. In some of them it produced material evidence of the Hebrew traditions and, with the help of characters that at first glance appeared to be molds for nails, science's perseverance read, in bricks of clay, the catastrophe of the Babylonian deluge. The magnificent subterranean temples of India, which an energetic race carved, as if it were the finest of laces, in the heart of the earth, and the imposing palaces that watch over Cochinchina, surrounded by its swampy solitude, were all described by England and France's indefatigable geniuses.

Recently science has revealed the immense sculpted stones boasted by Java and the enigmatic Khmer monuments, on which is drawn in relief—perhaps hundreds of centuries after it was devised—the death of a monkey king. Europe has unearthed massive constructions from early modern times, and explorations in the United States of America have yielded gigantic works of moved earth, a quiet testimony to the powerful and prolonged life of peoples long-since lost.

Discoveries that all went about stirring up controversies between the sacred and profane traditions. That held up as truths things previously believed to be fantasy, and that, more than anything, argued that an utterly remote past lay behind the civilization that moderns thought to be their own work, and which turned out to be the continuation of an ancient yet magnificent effort, whose power had run out in certain regions, in accordance with the law of creation that calls for everything, be it material, be it moral, to be born, live, and die.

And such has been the case with the progress of the races that make up humanity around the globe. They are born, they develop, they rise up, widen their sphere of action, link together and elaborate their faculties, but there comes a time when this growth is complete, decrepitude looms; they decay and disappear, victims of the struggle for existence, leaving their place to others, for whom this same evolutionary renewal starts once more.

This is how civilizations are born, progress, diminish, and go extinct; but later they come back on stage carrying new elements with them. This play is indestructible and only its appearance changes. It is the product of man's brain, unified as a genus, varied as a species.

All of these studies took into consideration civilized man; we have yet to find out what brought him to that point.

Thus, descending from the heights of ancient man's dazzling creations, it was necessary to go down to his humble cradle. The most recent excavations have shown, in some ancient cities, different superposed layers of rubbish, piled up by man and by time, that indicate different progressive civilizations. In each one of these cities it was possible to follow the evolution of a people, but the starting point was always an advanced state; nothing was revealed about the intervening period from the time nomadic life was abandoned to the adoption of stable dwellings.

The study of humankind demanded more; the geological layers that enclose the world's history, like the pages of a book, were still virgin to any investigation related to man; and he, with the unbreakable need and desire to trace his own footsteps from the remotest of ages, abandoning all worries, created prehistoric archaeology, which means the study of ancient things, offspring of his own creativity, prior to history, and he searched for a past that was more distant than history, traditions, or even legends would indicate.

It would have been impossible for man to have accomplished all that he did in just the short number of years that the sacred traditions allotted him to carry out his immense development.

He calmly broke down this barrier and set out to test with facts whether or not he had been a contemporary of the giant mammals, whose species, driven to extinction by cataclysms, had vanished. So there came the moment when, instead of believing in the remains of giant men, our genus's predecessors on this earth, and rejecting our own existence prior to the Bible's six thousand years, it was proven that man had lived at the same time as those giants, which were not humans but rather belonged to other genera, in eras prior to the most recent geological cataclysms, and that he had the same physical form that he has today in the midst of the colossi of creation, which his privileged constitution dominated and whose extinction he survived.

Just as the geologist is neither frightened by large numbers of centuries, nor counts time in years, but rather in millions of them in the most limited of cases, so was he unafraid to situate the beginning of human progress in epochs for which the duration of history is but a second, and which in turn are hardly noticeable in the immensity of the time that the earth has needed for all its transformations.

The geological periods in which man's existence is recorded—and which in relation to the present are so large that they have spawned incredulity as to this absolutely confirmed fact—are insignificant in comparison with the time needed to form just one of the many layers that make up a single epoch.

With all the strides in the field of knowledge that the masters of the globe were taking every day, there was an increasingly apparent gap in our knowledge, which these studies served to fill. A history that only reveals to us a handful of centuries, and which sets out, as I have already stated, from an advanced state of things, of which it itself is born, is unable to tell us all there is to know about the eras of the past, about which it can hardly even manage to stammer. As such, history had to make use of prehistoric archaeology in order to cast out the shadows surrounding the lost ages, a people's past. In order to make out the first steps of each people, the first lights provided by written history were insufficient. Egyptian papyrus scrolls and Assyrian cuneiform characters were silent in the face of man's great antiquity. When man's spirit felt the need to go all the way back to the times in which those lights did not yet exist, it had to look to geology and paleontology, sciences that lead it to ages so far back in time, that periods in the earth's history that were different from our own appeared. At the same time, it procured elements that repudiated any consanguineous relationship, in that period, between man and the pongids.

The Engis skull, extracted from the Quaternary mud by these sciences, and which, according to the saying of the learned Huxley, could have contained the brain of a philosopher or of a savage, and the constructions, product of man's intelligence, exhumed from that layer and even from the Tertiary, were to constitute insurmountable barriers for relations between the two genera.

We can count among this century's advancements, then, the reconstruction of the human genus's ante-historical past, sketched out in a diverse array of points that we have begun to link together.

However, though much has been done, the last word about our appearance on the earth has yet to be uttered, and observation shows that man's infancy was extremely long, that he began by groping about in the dark and evolved much like a child. A gradual growth of his faculties, a subsequent aim towards perfection, has given us history and civilization, which is his evolution brought to completion.

What we have learned is that in the beginning all men started out with the same brain; living under equal conditions he followed the same progressive pace. Anthropology and the sciences that it subsumes teach us that in all periods savages have employed the same methods and understood life in the same terms. In this way, the savage man living today in Chaco, in the Patagonian sands and in the frozen Tierra del Fuego, is the moral equal of the prehistoric inhabitant of Europe.

But there came a period in which certain species were given a stronger push, whose causes are still shrouded in mystery, and acquired, perhaps under the influence of the environment in which they lived, a moral and effective preponderance over the others, creating civilization while the rest continued to live as barbarians.

While it is impossible to comprehend a people's history when it begins at a point of advanced progress, it is equally obscure and problematic to look for the decisive causes behind one people's preponderance over another. This is why we must obtain, from all over the world, all the information we can that might shed light on the earth's inhabitants, along with data concerning these men's physiognomy, for this will allow us to make untold comparisons between existing and extinct populations' degrees of primitive civilization.

And so, joining together all precedents, seeking out man's anatomical and physiological conditions as well as his moral conditions, studying what he creates through his language and though his works, we will come, one day, to repair the progress of humanity, and we will find out why there are races whose progress has stagnated, and if they are morally condemned from childhood, or if their decrepitude is due to the influence of their surroundings.

In this way we will find out why the inhabitants of the torrid jungles, of the sands and of the frozen regions at either end of the New World, those of the territories that Livingston glorified with his presence, those of the Oceanic Islands, etc., have not followed the march of progress, and what ethnic conditions have condemned them.

All of the elements that the new sciences have contributed to the knowledge of the human genus have testified to the latter's presence at a time when the geographical and even geological physiognomy of the globe was different from the present, and when another fauna and another flora covered its surface.

Europe has pushed back the origin of its inhabitants to these times, and the vestiges that have been uncovered by archaeologists' picks point to types of humans that were different from those who presently inhabit these same regions and which are, on the contrary, similar to those who are today considered to be representatives of our genus's most inferior species. It has been shown that these types of men disappeared in a rapid extinction, assimilating into the ranks of the superior invaders or else leaving for other regions that offered them living conditions similar to the ones they left behind, due, among other reasons, to the absorbing force of the privileged types. These peoples disappeared, carrying with them the secret of their first homeland and of their emigrations. Those that left them behind or drove them to extinction later found themselves in more or less the same state, and today civilized men do not know the precise location of the encampments of their ancestors, who with time would go on to build entire nations.

There must not be a single place on the whole globe where a group of men inhabits the same ground on which their primitive ancestors began their evolution, whose first step, nomadic life, impedes such a thing, and studies would not likely be able to verify it. No autochthonous races are to be found alive today, and the geographical centers of creation will remain unknown to us for quite some time.

The first men's savage state, which provided them with no means of subsistence other than hunting, which in turn was incompatible with sedentary life, forced them to emigrate and gave rise to wars between the primitive tribes. The most abundant fishing and hunting grounds were the subject of the first feuds, as is still the case among the world's surviving savages. In this way, man's moral advancement, which got under way when he asserted himself over the wild beasts, gained ground as he exterminated his fellow men.

All men started out with the same level of intellect; but over time the struggle for an existence which was constantly under threat engendered differences. A contributing factor was the influence of the environments towards which their gradual development led them, promoting them to shepherds, then to farmers, along with the mixing with and absorption of superior types, who quenched the crude traits of their inferiors. In this way, men who belonging to latter group gradually disappeared.

The farther back in time that these events took place, and the greater the ethnic mixture that altered its composition, the higher the level of perfection attained by a people.

This is why certain races of the European type have attained a supremely high level of culture, far removed from the center where the extinct civilizations acquired the greater part of their strength, civilizations whose remnants came to form part of it, bringing with them their seeds and setting up camp where the land was still virgin to great works of genius.

It is not my intention to study this evolution that has given the world civilization, which has persisted through geological ages and which scholarship is now able to treat with a certain level of precision.

Since every people that takes an interest in inquiring into the origin of its prosperity and of its race attempts to trace, from period to period, the development of the men who have inhabited the land that it presently occupies as a nation, studying this development in its most insignificant manifestations, down to the nearly infantile stages when, in the beginning, men set about distancing themselves from the animal kingdom, and then up to the present day in which he has completely dominated it, from the age of the discovery of fire up to the age of steel, I only wish to demonstrate the interest, for the Argentine people, of the most remote past of our predecessors on this soil, as it is the foundation of our own history.

II

In America, as in Europe, the keenest of interests has been aroused in investigating the origin of the land's first inhabitants.

The more or less fantastical legends and traditions told by the first chroniclers, distorted by the period's religious spirit, took as fact the idea that America had been populated by the children of Noah and their first descendants. An intrinsically American race of man did not fit into their calculations; the Bible's lines of descent stood in the way of such an idea. And through these inexplicable aberrations some even went so far as to believe that the indigenous people of this land were not men, or were at least inferior to all the known varieties, and that he should therefore not be treated as one; perhaps it was believed that he was the offspring of one of the accursed races of the ancient legends.

And so, in the first period, entire nations, blood relatives of the Christian conquerors, were wiped out, annihilated by a war that was called holy; and not content with destroying the men, they tried to do away with their works as well.

The ignorant soldiery's greed melted down their metals, whose raw material was less precious than the artistic works it had been used to create, and fanaticism destroyed their temples—in many of which a faith more humane than that held by the invaders was preached—and burned their ancient manuscripts relating the history of America, virgin up to that point to such acts of vandalism.

A moment of moral failure, brought on by a mystical inebriation taken to the extreme, condemned America's past to darkness. The memory of how the works of the pagans were plundered, and of the clear vainglory in the chronicles of those responsible, causes indignation and grief. Only science can erase it, by using all its might to reconstruct the former splendor of the continent's masters, cleansing from those of us who descend from them, the stain that fell like a curse of the times upon our European forefathers, the conquerors.

Before long, the inhabitants of the American jungles, mountains, and pampas will have disappeared in the face of the annihilating power of modern civilization. These tribes, the remains of the ones that once dominated America, subject to the law of fate by which everything must go extinct when faced with a superior power, have fallen and returned to barbarism as they await the fulfillment of their destiny. Because they cannot rise up, pure, to the level of today's culture by means of any rapid transition, they will go extinct, and soon nothing will be left of them but their works and some of their traditions collected by those of us who have the good fortune of hearing them. We must not wait until that moment comes to trace the history of their time here on earth: the dust which will settle upon these poor tribes will not allow us, save with great efforts, to decipher, and even then only in part, the secrets that they will take with them when they finally succumb.

The time when it was believed that American riches had once helped build King Solomon's treasure, that the seafaring nations of the Old World had been in contact since the most remote antiquity with the New, are slowly being forgotten in the face of modern discoveries.

Anthropology has begun to take up American man, as one of the most important problems that lay before it.

A theory exists that America was initially populated by foreign migrations, perhaps Polynesian, Buddhist, Phoenician, Egyptian, or Scandinavian, from which descended the great civilizations of México and Perú, which would have taken place in more or less remote times, but in any case not too long before the historical period, and would consequently form part of the present era. This has, however, run into many obstacles that keep it from being adequately explained.

As long as we had no information about the Americans' antiquity other than these civilizations, this theory could be defended. Yet today, with our new studies and discoveries, the idea that America was populated by races with a highly developed culture is losing its foundation, which rested upon the lack of information concerning ante-historical times.

These chronologies, more or less hypothetical, attributed to American civilization but a few short centuries, and in that period of time they tried to fit what would have called for tens of thousands of them. To allot only a few hundred years to the development of Peru is simply ridiculous, and yet, for years this course of time has been assumed for the development of the Incan Empire's immense prosperity. The little-known traditions naïvely related by Montesinos, who gives them little credibility, situate their beginnings several centuries farther in the past; but this is still insignificant next to the time needed for the unfolding of civilizations on par with those that were played out across the Bolivian highlands and the Peruvian coasts.

The discovery of the remnants of humans both ancient and modern complicates the issue of American man to such an extent that the system of classification used up till now is finally being discarded. As the light shines brighter, we can begin to make out through the shadows that cloak our past, objects of study that are so new that they would dumbfound anyone who attempted to orient himself inside such a labyrinth of unknown elements, the sheer quantity of which will ensure a long-lasting reign of chaos surrounding the highly important subject of our origins.

The prototype that the learned Mr. Morton considered to be the ethnic core of all the races dispersed across the American territory is impossible to accept in light of the recent discoveries that have dealt a fatal blow to the theory of the human species' unity. The presence of human fossils and of heretofore unfamiliar races has provided the opposing theory with significant reinforcements. Monogenesis is fading, and polygenesis has an increasingly high probability of being correct in its calculations.

The supporters of the monogenist theory have made every possible effort in order to explain how America was populated by means of migrations, bringing to the table a great quantity of data that, while they may shed light on the contact carried on in ancient times between our continent and Asia, lose their importance following the discovery of men who witnessed the geological events that gave these regions their present shape and who appear to be the contemporaries of the great extinct mammals.

Rather, the discovery in the United States, México, and Brazil of the fossilized remains of men from the Quaternary period serves to encourage those of us who defend the existence of an autochthonous American man, whose anthropological traits are yet to be determined with certainty; and the studies carried out in these same countries, along with Central America, New Granada, Venezuela, Guiana, Perú, Chile, and our own Republic demonstrate the multiplicity of the American races.

Thus, the physical and moral history of the Americans, whose origin was once discussed only in legends, grows in interest with each passing day.

If God created flies in America, why wouldn't he have created men? asked Voltaire; and science is going to respond to the great French genius's conjecture with the testimony of truth.

The autochthonism of the American races will be proven before long, and we will be able to say with confidence that they did not descend from Europeans, Asians, Africans, or Polynesians, but rather from Americans, admitting nevertheless the hypothesis that the pre-Columbian social advancement of our indigenous predecessors was the product of convergence with Asians and Polynesians.

These explorations still do not allow us to paint one continuous picture of America's ante-history. Their findings, though of the utmost importance, are too isolated for us to even make a sketch of it. For now, the Americanologist should restrict his mission to the gathering of data that can eventually make up a general schema. Furthermore, I believe that research into the past of all the many nations of the Americas, in particular those of the boreal regions, should not yet form part of the course of study that falls upon those of us who live in this part of the continent to undertake. It is too complicated and we are too far removed from these areas to be able to deal with the matter in any sort of useful fashion. In my opinion we should wait until each of the two Americas has been completely studied on its own, at which point we should we join together at once all of the data we have obtained, if science is to show us, as we can only expect that it will, the contact between and migrations of the races that inhabited these two great divisions of the New World.

There is no doubt that the mixing of a number of races gave rise to the present-day civilization in the other hemisphere, the civilization that is currently taking hold in our own hemisphere, where human intelligence has before it a marvelous future and perhaps the future center of progress as well, once the old lands grow sterile from overproduction, and the extinct civilizations have ceased to renew their ethnic elements and disappear. The stagnant civilizations of Asia and Egypt were extinguished due to this same cause, which deprived them of a necessary ingredient for life, just as a lamp is extinguished when it runs out of oil, shooting off sparks just before it burns out. Some of these sparks were put to use by the races that were heading towards progress; Greece was born of them, and has continued to spread them up to the present day. And yet America was too far away to take advantage of those spoils that would have given it a new life then and there.

In the fifteenth century its own period of decrepitude had arrived, and its decaying civilization was on the point of burning out before it could be carried on through assimilation with foreigners, who rather preferred to hasten its demise.

All the data indicate that the first manifestations of intelligence on the American continent were the same as in the Old World; this is made clear by the methods used to fashion primitive weapons, which surely were instinctive, rather than revealed.

The arrival of new races, Asian and Oceanic, who already possessed relatively superior intellectual faculties, provided the mixing that was necessary in order to move toward perfection. The mutual assimilation of these races slowly engendered the social development, in different territories and under distinct circumstances, that resulted in the marvelous civilizations of México, Central America, Perú, and Bolivia. Since these ethnic elements were still insufficient for evolution to continue, the era of decadence arrived, as in Egypt, Assyria, and India. For these reasons I believe that the immense ruins that can be admired all across the American territory, though they arose out of primitive and diverse racial elements, both American and foreign, are nonetheless testimonies to human intelligence, which was developed in America at the same time that the now lost civilizations of the Old World were also beginning their development. Perhaps many of these great ruins belong to peoples whose period of decadence began at the same time as that of the peoples who erected the Hindu temples, the Assyrian palaces, and the Egyptian pyramids.

The civilization that the Spanish found in Perú was another, more modern one—though it was also in decline because it lacked the means that induce social transformations—and one that in turn rested upon other, more ancient civilizations.

It is evident, then, that the migrations to America did not bring with them a pre-made civilization, nor did they establish the first population center, as many would have it. We also must infer that these migrations were made up of men who belonged to still uncultivated races. However, it is equally certain that those who believe that this civilization resulted exclusively from the natural development of the Americans' own spirit, and that no other foreign races had any hand in it, also set out from a flawed premise. There is not a single example of a people or a race that rose to the heights of civilization all on its own, and without a doubt they have always relied on the incorporation of other ones, more or less superior, though perhaps still uncultivated, in order to achieve it.

We Americans should not ease our efforts until we have thoroughly dispelled all the shadows from this picture. Let us use the immense collection of materials that we can have at our disposal to probe the history of our land's indigenous people. Yet before we study how the peoples that made up the great American family progressed in their civilization, let us first discover that civilization's origin, and may this serve as our starting point.

Let us scour the geological layers and extract the earliest remnants; these will show us an anthropological prototype identical to many of Europe's first inhabitants. Some of the skulls will exhibit pongid characteristics. The products of their handiwork will also be the same. Let us study the tribes that still live as savages and compare them with the results of these remnants, and we will find infinite analogies which might allow us to reconstruct the history of our fossil grandfathers.

Let us scour the mounds of earth that can be seen by the seashores and the riverbanks, and we will be able to observe the handiwork of the fishing or ichthyophagous tribes all along the American coastline, one similar to that of the prehistoric fishermen of Scandinavia, which will raise the level of culture conceded to our predecessors. Comparing the objects that are extracted from these mounds with those that are used by the present-day inhabitants of part of the West Patagonian islands, we will reconstruct their domestic life and their customs.

Let us continue onward and we will find, here in America, Tumuli, Cromlechs, and Dolmens equal to those of Europe, Asia, and Africa; similar monoliths; and the stone age will yield an immense body of material for study, from the Patagonian sands to warm Guiana. Linking all of this together we will sketch out a civilization whose progress followed a pace equal to that of the Old World.

Let us scour the mountain caves and we will discover ancient human dwellings like those still built today by the Patagons; let us search among the rocks that form their walls, or in the free-standing stones on the riverbanks, or in the middle of the pampas or the forests, and in surprise we will behold carved and painted signs or characters that our current knowledge is unable to interpret.

Patient investigation will come to show us that the signs that so surprised the illustrious Mr. Humboldt and revealed to him the existence of a great people, ancient and extinct, in the midst of those lush jungles and next to the Orinoco's craggy waterfalls, are not landlocked by hundreds of leagues of surface, but rather by dozens of thousands. It will show us that, with few differences, the same signs are found all across America, from the islands of Vancouver near the Boreal circle, to the Argentine lakes of Patagonia, and that the figures painted there on the rugged, vertical walls of Punta Walichu are nearly identical to those found by explorers from the United States in Arizona, to the north of Mexico; that the ancient stone carvings distributed across México, Central America, Guiana, Brazil, Perú, Bolivia, Chile, and the Argentine Republic all appear to be the work of a single race. I am convinced that craniology, with the help of archaeology, is going to show us that this was what is now known as the ancient Cariban race, which will be found to be of the same ilk as the deformed macrocephalic skulls found all the way from Isla de Sacrificios in the Gulf of Mexico down to Patagonia, along with those that travelers have extracted from the Bolivian necropolises, attributed due to lack of studies to the builders of the monolithic constructions at Tiahuanaco, christened with the name Aymaraes.

The American anthropologist's field of study is wide open, and a mere list of the transcendental questions that he has been called upon to answer is long indeed. What race sealed up the remains of their dead in funerary urns that are found in both of the Americas, and which was not the cited Cariban race, as is demonstrated by the skulls that are preserved in this very city's Museum of Anthropology? We will find that this race lived in Brazil, represented by the ancient Coroado, among others, in Peru prior to the time of the Incas, in the Andean foothills, in the valleys of the Calchaquí, in the sands of Chaco, in Paraguay, on the islands of the Paraná River, and in the Oriental Republic of Uruguay.

Who were the Atumurumas, the ancient moon-worshipping inhabitants of Bolivia? Who were the Chimú, whose magnificent works of art were studied by the learned Mr. Squier during his exploration of Peru, and whose activity, evidence of which is preserved on the volcanic slopes of the Andes, is beyond all compare?

Who were the other Peruvian races, who still remain largely unknown? Who were the Quechua, from among whom was born the Incaic dynasty that in its period of preeminence would send out its chiefs, new Alexanders, to conquer remote countries that the descendents of Europeans have still yet to reach, constructing roads thousands of leagues in length and climbing on foot, time and again, those Andes from which they came. Who were the men who inhabited Ecuador, the Chibcha or the Muisca who worked with gold and precious stones? Our studies will reveal the answers to us in time, and will dispel all doubts surrounding these points, and those concerning the rest of the tribes that are still alive today or that have recently disappeared from the rest of the continent, and that I have not mentioned. One day we will also learn about the life of the brave and obscure Charrúa, the Querandí, and the race that turned southern Chile into a new Sparta.

My good fortune has provided me with numerous elements that have helped to cast a fair amount of light on America's origins. Prompted by the interest of science alone, I would like to say that my theory—of an initial primitive inhabitant of America and a later fusion with Asians and Polynesians in ante-historical times that resulted, by overcoming that initial element, in civilization—is slowly being accepted.

The learned Dr. Topinard, in his new book on anthropology, has made a distinction, based on the skulls in the collection that I assembled in Patagonia, between the Patagon and the American—which is to say, the present-day American who was believed to comprise just one unified type—and has established another entirely separate type. In a letter dated June 15th of last year, he wrote, among other paragraphs: "I see nothing more interesting on anthropology's horizon than this great discovery (. . .) This idea of a dolichocephalic race, autochthonous to America and whose type would be of the sort of your fathers, must necessarily be introduced to the public under your name as soon as possible."

If some good fortune has favored me, and grants me the facts that will allow me to prove this already accepted theory, that there was once an intrinsically American man, then we must continue the studies that we have already begun; and since there is no need to restore to America the discovery of its primitive man's physical structure, almost identical to that of the ancient European, let us find out what course he followed from the time of his origin.

Let us use the study of languages to find which of them are autochthonous.

Let us study myths and religions, and we will see that Americans' beliefs have followed the same course as those of the Old World, as evinced by the appearance of moon worship in Bolivia's remotest past, which Dr. D. Vicente F. López brought to light years ago, and which has recently been confirmed by Squier and is also depicted on an ancient cup preserved in the Museum of Anthropology, a donation of General D. Bartholomé Mitre.

Twelve years have passed since I began the collections that have been used as a foundation for the Museum of Anthropology and Archaeology of Buenos Aires, which has just been opened for the public to study.

The collections are modest, a reminder of their origin, but together form a nursery bed out of which will rise the scientific monument to commemorate and safeguard the physical history of the indigenous Americans, before their mixing with Europeans, and to house the products of their handiwork, humble in the beginning and then magnificent, which we have trampled upon indifferently up till now.

Let us try to build on them: in our vast territory we have all the desirable elements needed to discover our origins and our advances, from the most barbarous tribes to the most civilized and now extinct ones, and by bringing them together we will be able to pass on the memory of our indigenous predecessors.

There is an immense museum sitting in the superficial layers of this Republic's soil: let us bring it to light. Let us classify it and exhibit it in an appropriate venue, where it might be displayed so as to arouse the imagination; and then today's American will reconstruct the domestic and warrior life of the pre-Columbian Americans with passable accuracy. Let us gather together inside of it all of the records that exist, whether printed, hand-written, or oral, in the form of traditions or legends, and make use of them to reconstruct our past.

I can assure you that our spirits will experience unrivaled moments of pleasure when they are carried back to the eras in which our ancestors ate wild fruits, and when they compare the immense evolution that has been carried out in their intellectual disposition. The viewer's eye will see there, in that retrospective museum, a whole series of social epochs and the remains of the men that shaped them. It will embrace everything from the time of the first sheets of flint, to the glorious deeds of the Charrúa, the Querandí, the Araucanians, and the Calchaquí.

I have tried to show, in what I have just written, the importance of the studies that should be undertaken by the museum, which opens to the public on August 1st thanks to the efficient support of the members of its *Society of Sponsors*. To conclude, let me express a desire that I hold, that on of the anniversary of this city's tricentennial, which will be marked with an exhibition of American industrial advancements of European origin, we might celebrate doubly on that day by exhibiting all that has been preserved of Argentina's physical and moral history, prior to this city's founding and to the discovery of America.

Francisco P. Moreno, Buenos Aires, July 30th, 1878

NOTE

1. TN: Moreno uses both *prehistórico/prehistoria* and *ante-histórico/ante-historia* to describe the period prior to the historical record. As such, I have chosen to respect what is perhaps a term of his own coinage, and translate the former as "prehistorical/prehistory" and the latter as "ante-historical/ante-history."

DARWIN
Lecture given in the Teatro Nacional, following the death of Charles Darwin

(May 30, 1882)

Domingo F. Sarmiento

Ladies and Gentlemen:

I have been invited by the Medical Circle to pronounce, in its name, some solemn words of respect and admiration for one of the greatest contemporary thinkers, for the most penetrating observer, for the most reflective and calm innovator, for the most humble and honorable expounder, in a word, for Darwin, who at the age of eighty-three passed on from a life of labor, after providing science with ever more insightful books in his final years, as if he were afraid of carrying with him to the grave the secret of his final studies, despite already having permeated the century with his name.

As I make this commemoration, I keep in mind that the beginnings of his illustrious career were these Argentine pampas and that Strait of Magellan and Tierra del Fuego that he explored, and so I can rest assured that those who do me the honor of listening to me will be indulgent; and that he will have the sympathy of the ladies present, if I add that Darwin ended his long life of labor surrounded by his family, which was raised just as he was in the simplicity of English country life, which is as comfortable as it is intelligent.

This noble display in honor of one of the greatest talents of our times should not surprise anyone, since great names that figure in the annals of scientific progress honor us frequently enough by binding themselves to our history and to our own progress as well.

Among them, in the forefront, stands the engineer Azara, who, spurred by the abundance of his collections, invented a system of classification for birds and quadrupeds, which turned out to be essentially a repetition of the Linnaean system. Bonpland was the first French emigrant to penetrate this American continent with Humboldt, and he stayed on until his death in

Corrientes. D'Orbigny preceded Bravard, and both of them described the pampas, postulating for it a different geological origin. Today this is known by the name of Pampean formation, the last page of creation, covered in hieroglyphics that our compatriots have learned to translate, and like the Egyptian fellahin with their pharaohs and mummies, they sell Megatheria, Clyptodons, Mylodons, and ancient horses that must not have heeded the call, since they were not saved on Noah's ark.

And finally, amongst us we have the learned Mr. Burmeister, the first paleontologist, who wrote in our homeland his *History of Creation*, whilst the wise Mr. Gould was preparing the final corrected and expanded edition of *The Skies*,[1] at our Observatory in Córdoba.

Why shouldn't we join the rest of the world in paying homage to the memory of Darwin, if the traces of his route through our territory are still fresh, if he is one of our very own scholars?

Assigned to speak about things having to do with the natural sciences and with the doctrines of a great naturalist, I feel more at ease with the members of the Medical Circle than with the large crowd that is perhaps hoping to hear from my lips some enlightening exposition of the ideas that have made Darwin the cornerstone of scandal insofar as the origin and descent of man are concerned. But the young physicians who are initiated in the sciences that go along with their profession know what to abide by in this respect, and what they know best is that I lack the authority to give an opinion on matters that stray from, or have not yet entered into, the sphere of public life, which has been my particular terrain.

I will not stray, then, from my habitual grounds.

I could say, gentlemen, that forty years ago I was already familiar with Darwin's name when, aboard Fitz-Roy's *Beagle*, he visited the southern tip of the continent, because I met the boat and its crew, and of course saw the diary, *Voyage of the Beagle*, which I would later cite on more than one occasion when speaking of the Strait. You will recall that I never was very jealous over our austral possessions, because I did not deem their defense worthy of a single barrel of gunpowder, and reproved the use of fantastical descriptions to bolster the imagination of these peoples who still hope to find *El Dorado*, sought out in vain by our fathers in those same regions, in an attempt to avoid a war in the name of that Holy Sepulcher of traditional illusions.

II

I would not dare to put forth any opinion of my own about the basis of Darwin's theory in the presence of my illustrious friend, the learned Mr. Burmeister, who does not accept it as a proven natural system, rejecting it for being the effect a non-scientific method, insomuch as it does not follow from recognized and undisputed facts, in order to rise above their existence to the cause that produces them.

Ten years ago the naturalist Agassiz, who carried out such penetrating studies on the glacial system, was of the same opinion; and apart from what I heard him say to this effect in Cambridge, he declared the same in precise terms in a printed lecture. "They asked me," he said, "what goal drove me to set out to explore the Amazon River: naturally, to enlarge my collection of fish for my studies; but the real issue that dragged down there was the hope that I could demonstrate that the theory of transformism was not based on facts."

In another part he lays down his doctrine clearly, saying, "all the derivations known to us are monstrosities, and the occurrence of these, under disturbing influences, are to my mind only additional evidence of the fixity of species."[2]

I should add, so that his latter declarations might be better appreciated, that, finding the idea of the transformation of species to be unfounded, "I am satisfied that, unless it can be shown that the differences between the Indian, negro, and white races are unstable and transient, it is not in keeping with the facts to affirm a community of origin for all the varieties of the human family, nor in keeping with scientific principles to make a difference between human races and animal species in a systematic point of view."[3]

A learned man of the caliber of Agassiz, whose work rests on the scientific theory of the separate origin of races, is not exactly predisposed to follow the first indication to the contrary.

Upon seeing Indians and blacks, however, he cannot resist the temptation to compare the former with the structure of the baboon, and the latter with that of the chimpanzee.

Nevertheless, he ends his expedition with marvelous results, and in a letter to the Emperor of Brazil, thanking him for the generous protection with which he has provided Agassiz's enterprise, he adds the following commentary, which, as it is both brief and grand, may be repeated before our audience, and all the more so because it is to be found in a book written by the wife of Agassiz, which will arouse the interest of the women present in a work written by a lady.

"I estimate at one thousand eight hundred," he says, "the *fish species* that I possess at present, and they will surely reach two thousand. I will not insist on how surprising this variety of fish species is in the waters of the Amazon and tributaries, though only with difficulty can I assimilate the idea that the Amazon supports twice as many species as the Mediterranean and a greater number than all the Atlantic from pole to pole.

"But it is not only the number of species that will surprise the naturalists. The fact that the greater part are circumscribed within small areas is still more surprising; AND THIS WILL CERTAINLY HAVE A DIRECT INFLUENCE UPON THE IDEAS CIRCULATING AT PRESENT ABOUT THE ORIGIN OF LIVING CREATURES."

I do not recall any of the supporters of transformism having known of this declaration by Agassiz, and I do not know whether he adopted it sincerely later on, just as Lyell accepted the existence of the human fossil, after having denied it for twenty years.

The Amazon has at least six hundred tributary rivers, almost all of them navigable, and in each one there are three divisions among the types of fish species, ones that live at the river's mouth, others in the middle part, and others near the source, without mixing amongst themselves, in addition to other species that travel along the entire length of the Amazon and up its tributaries. One could say that these latter ones are the government of the Amazonian nation, and the others constitute its provincialists.

We would have to suppose that the Creator awoke in a very good mood on the fifth day, and looked upon Brazil with special favor, in order to put, all in one sitting, one thousand eight hundred different species of fish in the Amazon river, fish so well behaved that they remain up to the present day in the spots assigned to each of them.

Darwin has simplified the task with his explanation of the variability of living creatures in accordance with their needs and location. It is well known that it was a little bird, Chilean in origin, that he found on the Galapagos archipelago, which aroused his suspicions. For, while remaining essentially the same, it had modified its beak to be short, long, thick, or thin depending on whether insects, seeds, grain, or hard nuts were to be found in its particular environs. Similarly the eucalyptus, practically the only type of tree in all Australia, and which is so familiar to us, has acquired hundreds of forms depending on whether the terrain is swampy or dry, valley or mountain.

Let us content ourselves, then, with that quasi-confession of the man who was dragged all the way to the Amazon in hopes of discovering evidence, which he did not find, to oppose the idea of transformism, so that we who settle for fewer species of fish in our rivers—and who can count in our favor the opinion of our own paleontologist and naturalist from Mercedes, Mr. Ameghino, who holds the same opinions as Darwin about the terrain that he covered—might not feel too ashamed to believe that each of us present here was once a monkey . . . and what lovely monkeys indeed!

And there are among us many who, for their own reasons, believe, practice, and test the doctrines of the illustrious thinker, and who just so happen to grow rich off of their belief—which is something that happens to each of us who believes in human progress.

Intelligent breeders of sheep are perfect and unrivaled Darwinists in the art of *varying species*.

It was from them that Darwin took his first notions, right here, in our very pastures, notions that he went on to perfect by dedicating himself to raising pigeons, which in Europe is the art of creating new varieties at the will of the breeder's fancy.

It was also here that he saw the colts with bands on their legs, which appear to point to the domestic horse's line of descent, or its relation to the zebra or quagga, bands that disappear later on in life.

In our country there are hundreds of farmers, breeders of both sheep and other animals. Among them the most prominent are the Pereira, Duportal, Chás, Ocampo, Olivera, Casares, Kemmis, and Lowry families, who read Darwin through and through, down to his periods and commas, wherever he speaks of variation through natural selection, since they themselves practice artificial selection by choosing the animals that will go on to reproduce. As for the rest of it, they don't care a farthing that the master himself descends from another cross-breeding and another selection.

We have given, then, science and fame to Darwin, with Argentine fossils and stock; and following his directions our farmers grow rich.

It seems to me that we Argentineans have sufficient reason to support the transformist doctrine, since we ourselves have transformed one variety of sheep into another. We have created a new species: the *argentiferous sheep*, named thus because its owner earns plenty of silver and because it is from Argentina as well.

III

Since you have asked me to give you an idea of what the new theory consists of and for what reason so many are so irked by it; and since those listening to me have no more faith in my aptitude for natural science than the little faith I myself have, I will use a homespun explanation that the learned Mr. Huxley gave in London, in a lecture before both gentlemen and ladies, to explain this very concept.

"The researches of the last three-quarters of a century have, in truth, revealed a wonderful richness of organic life in those rocks. Certainly not fewer than thirty or forty thousand different species of fossils have been discovered. You have no more ground for doubting that these creatures really lived and died at or near the places in which we find them than you have for like scepticism about a shell on the sea-shore. [. . .]

"[T]he first point for us is to examine how much the extinct 'Flora' and 'Fauna' . . . differ from the 'Flora' and 'Fauna' of the present day. [. . .]

"If you divide the animal kingdom into orders, you will find that there are about one hundred and twenty. [. . .]

"Now, how many of those are absolutely extinct? [. . .]

"Among the mammalia and birds there are none extinct; [. . .] in the reptile class, there are no less than half of the orders which are absolutely extinct. If we turn to the 'Amphibia,' there was one extinct order, [. . .]

"No order of fishes is known to be extinct. [. . .]

"There is not known to be a single ordinal form of insect extinct. There are only two orders extinct among the 'Crustacea.' There is not known to be an extinct order of these creatures, the parasitic and other worms; but there are two, not to say three, absolutely extinct orders of this class, the 'Echinodermata'; out of all the orders of the 'Coelenterata' and 'Protozoa' only one, [. . .] [s]o that, you see, out of somewhere about 120 orders of animals, taking them altogether, you will not, at the outside estimate, find above ten or a dozen extinct."

Now, in term of succession, Huxley, who I am citing, graphically exemplifies it thus:

"Suppose that we were to sink a vertical pit through the floor beneath us, and that I could succeed in making a section right through in the direction of New Zealand, I should find in each of the different beds through which I passed the remains of animals which I should find in that stratum and not in the others. First, I should come upon beds of gravel or drift containing the bones of large animals, such as the elephant, rhinoceros, and cave tiger. Rather curious things to fall across in Piccadilly! If I should dig lower still, I should come upon a bed of what we call the London clay, and in this, as you will see in our galleries upstairs, are found remains of strange cattle, remains of turtles, palms, and large tropical fruits; with shell-fish such as you see the like of now only in tropical regions. If I went below that, I should come upon the chalk, and there I should find something altogether different, the remains of ichthyosauri and pterodactyles, and ammonites, and so forth. [. . .]

"[I]nasmuch as we started with the clear principle, that in a series of naturally-disposed mud beds the lowest are the oldest, we should come to this result, that the further we go back in time the more difference exists between the animal and vegetable life of an epoch and that which now exists. [. . .]

"[I]f you go through the enormous thickness of the earth's crust and get down to the older rocks, the higher vertebrate animals—the quadrupeds, birds, and fishes—cease to be found; beneath them you find only the invertebrate animals; and in the deepest and lowest rocks those remains become scantier and scantier, not in any very gradual progression, however, until, at length, in what are supposed to be the oldest rocks, the animal remains which are found are almost always confined to four forms—'Oldhamia,' whose precise nature is not known, whether plant or animal [. . .] and three undoubted animals—two crustaceans and one mollusc." This is the most concise notion to which paleontology, or the creation of the animals, can be reduced by examination of its remains.[4] To see with our own eyes the evidence of these phenomena, all we need to do is have a look at the Museum of Buenos Aires, which is a genuine cemetery of past creations.

What has happened, then, in this branch of natural history, is the same as what happened in astronomy. The plants and animals divided up by Linnaeus and Buffon into genera, subgenera, species, families, varieties—What a wonderful thing! What an orderly and wise creation!—.[5] But Cuvier was creating comparative anatomy, and in the meantime, the bones of unknown animals were being unearthed all across Europe, and it turns out that there were elephants, rhinoceroses, and hippopotamuses beneath London; and beneath Paris, elk like we find in Palermo; but these are not the same animals that live in Africa today, nor in India. It turns out that further down, in another deposit, there was another elephant that was simpler than that of the present day, such that the young of today's elephant are comparable to the adults of the ancient variety. This is a law that Agassiz has observed in palm trees, the sapling of one species being the pattern for the grown tree of another inferior species in the inverse order of succession; and we have all observed that the eucalyptus has one form when it sprouts, and takes on a different form at a certain point in its development.

It turns out that the animals were not all created at once, but rather thousands of centuries passed between the different deposits; and, for example, today's elephant is not the same as the one that was created three or four times prior with less perfect forms.

Comparative anatomy revealed yet another fact, which is that all mammals share the same prototype, translated in various forms, according to whether the creature in question is a man, dog, bird, or turtle: a backbone, a neck, four legs, ending, even in the wings of the bird, in three, four, or five fingers.

Embryology has discovered the same phenomenon in the different stages of the human fetus as in the great mass of animal creation.

All embryos come from an egg, incubated inside or outside the body, according to how close the animal has come to approaching perfection; it is at first a mass, as with the mollusks, then an embryo, with the same form for dog, man, cat, until at a certain point each one continues along its own path, let us say, and is differentiated according to its genus, and moves along in its development, to become a fish, or a breathing mammal, or perhaps a biped . . . QUE NOUS VOICI! . . .

These are all indisputable facts, which no one would dare question today without placing himself in conflict with science itself.

And here we run into the once dominant ideas about creation—which most certainly do not have to do with the facts—and the human spirit again finds itself uneasy, dismantled, and without foundation.

Chaos has been let loose, and a law is needed to explain and bring together the twenty-six creations postulated by Elie de Beaumont, carried out in hardly a million years.

The naïve illusions of the primitive peoples are worth examining.

Do you know, gentlemen, how it rains?

Well neither do I; and our primitive ancestors all the less so.

How advanced the sciences must be, in order to figure out how water gathers in the sky and from time to time falls in torrents, as if rivers cascaded down from above!

I received my first notions of the rain from an eleven-year-old girl when I myself was even younger than her.

White clouds sailed across the blue sky one summer morning, and the little girl observed as she watched them:

"They are going to the ocean to get water."

I never forgot that. In San Juan, at the feet of the Andes, the ocean is unknown. If an ignorant boy, who doesn't even know how to read, the child of ignorant parents, says the word *ocean* it is only because it exists in the Spanish language, brought over by the European settlers who sailed across it. And so I completed my theory of the rain. At that moment I understood as clear as the light of day where the clouds got their water: they brought it from the ocean! And, what a surprise, all of you know that this is exactly the case! This is the water cycle, upon which I will not dwell.

But it must have taken primitive man thousands of centuries to understand where the clouds got their water; just as they would never have understood how the planets were suspended in the sky.

For him the earth was held up by pillars on the back of a turtle, the turtle swam across the abyss—and now please don't go on, sir.

But science has a different way of explaining things.

In the beginning the inchoate light was diffuse, as can still be seen in the Orion nebula and in the irreducible nebulas of star dust in the Milky Way. The light contained the material that yields all the bands of the spectrum, and, rending itself apart, formed nebulas that began to rotate due to the molecules' gravitational pull. As they condensed into suns like our own, they created, through centrifugal force, equatorial rings like those we find still unbroken on Saturn, and which in other cases have broken up to form the planets Neptune, Uranus, and Jupiter, which stand like cosmic surveying poles to mark the space occupied eons ago by the nebulous sun; and given that there are seventy million stars, which are really so many suns, there are seventy million centers of creation just like our own.

Newton put these worlds in order, giving them laws; Laplace and Herschell have described their lines of succession and their development. Mr. Gould is on the lookout for signs of how these laws are put into effect and for new developments in those seemingly unchangeable worlds that, in reality, are eternally in motion.

And now we come to the earth, where it would seem that on an infinitesimally small scale the same course of events has taken place. First it was a scrap or a spark that escaped from the solar forge. It burned for a time; it slowly cooled down; the metallic gases condensed into liquids at the center of the ball that formed as the mass spun around on its axis. Layers of lime, salt, clay, etc., formed around this center, until there was an outer crust that allowed the water vapor to condense and form clouds, which fell to the surface and formed warm seas, out of which islands emerged. And as the crust underwent a continual fluctuation and break-up, the seas evaporated, fell to the earth as water again, lands emerged, lichens appeared, then ferns, palm trees, conifers, deciduous trees, mono- and dicotyledons, down to the present day. And at the same time, shoals of *monera*, living matter lacking organs, appeared in the sea, and then *amoebas*, living matter organized around a single nucleus, and then crustaceans, mollusks, fish, amphibians, quadrupeds and quadrumanous man-like animals, and the last to appear on the earth, those of us who now stand on two feet, having long ago stopped living in the trees like the natives of Australia do up to this day in their eucalyptus trees. It is Darwin who says this.

IV

I will allow myself to continue along this retrospective course, searching out a formula, like the one that Newton found after Copernicus and Galileo had put things in their place.

Might there not be some law that could link together, in just one element, all of these scattered creations over the course of millions of years, replacing some here with others there, introducing more advanced forms following from the same prototype, right up to the appearance of man, who is similar to the monkey, which is similar to the quadrupeds—which is what the *lemur* is, in fact—which look like certain amphibians, which are much like fish, which are not so far off from a worm that is found on certain beaches and, in its larval state, grows a dorsal spine, which is what constitutes the barrier between mollusks and crustaceans on the one hand, and more advanced forms of creation on the other?

All of you are doing just what Darwin himself did after having crossed the seas and examined the transformation undergone by that little Chilean bird, or those fantastical transformations that he carried out on his pigeons, extending their beaks, shortening them, all in accordance with the whims of their breeder.

I have still not completed the series of movements that begin with the diffuse light that still forms part of the Orion nebula.

We now have before us man, a variation on our ancestor the monkey, a certain dandy back down the bloodline of our family, the hominids, one Mr. Gorilla or what have you; but he is still unarmed, naked, and furthermore mute from birth.

His story began long before now, some five or six thousand years ago with the Hebrews—great knowers of the Divine—with the Egyptians—great builders of pyramids—with the Greeks—great lovers of beauty. The Romans—well, they are essentially us.

But there was a page missing from human history, one that Columbus discovered in America, here on our soil, namely *primitive* man, with no art, with no iron or bronze with which to make weapons, regardless of whether in some spots he knew of gold and silver, wove fibers, and built temples. Here, in our country, in the pampas and Patagonia, all he knew was fire, with no greater weapons than little pieces of stone to skin llamas and scrape their hides.

The Indian Manuel Grande built himself a little yard-high enclosure out of sticks, when he was imprisoned by the Government on the isle of Martín García, and there he lived for four months with eight strapping young lads from his tribe. Gorillas do the same thing.

In Aurignac archeologists have found the tomb of a primitive man was who has restored the first page of human history and shown man his origins. His story is much the same. He was a dull animal, itinerant, seeking refuge in caverns, the contemporary of two or three extinct animal species.

He lived without any weapons other than pieces of flint that fit in the palm of his hand, as if in the beginning man felt he had to fit himself with hands of stone, in the absence of the claws and horns that his enemies possessed and he envied.

Later on he made spearheads, knives, awls, and various other tools that I do not know how to describe, but which Mr. Ameghino has arranged in order at the Exhibition, according to their time period, their use, and the progresses of that "literature" from before there was learning, from the eras of coarse stone to the eras of polished stone.

Through the study of and comparison with what has already been established in Europe, Mr. Ameghino has continued to advance the known history of Argentina and America's prehistoric man, who is but the prologue to the more advanced history of the man who used bronze to make weapons for combat, and pottery to provide himself with cups, and on to iron, which I believe was discovered by the blacks, who to this day still smelt it using the most primitive of methods, shaping it in different ways. Yet still they have not become civilized, despite what the primitive race, that first sketch of mankind, believes. They live together to this day with *gorillas*—since the *orang* did not make it off the far-away island of Java, which is believed to be the peak of mountains that were submerged along with the lands where man first appeared, perhaps part of the race of Negroes that live on in Australia.

I suppose that you, the ladies here present, would not want to be descendents of the Negroes of Fiji either, who eat their mothers and adorn their heads with such elaborate coiffures that it takes the hairdresser three days to erect that majestic edifice, and the dandy who dons it sleeps with his neck propped up in the air on a wooden fork so as not to rumple the coiffure, which should last six or seven days. This manner of pillow is yet another trinket from the primitive eras of man.

Was man born with the ability to talk?

Of women some noteworthy authors believe it possible.

Nevertheless, a woman in New York once married a Chinaman who did not speak English and who, when filing a suit against him for misbehavior, was asked by the judge: "How could the two of you even communicate to each other that you wanted to marry?" "Well, the thing is," replied the lady, "there's no shortage of ways to communicate that."

The same thing happened with languages as with astronomy, natural history, and human history.

At first, incoherent facts that didn't follow from one general idea: Max-Müller has settled this matter.

Just as man went about inventing stone weapons for himself, he invented three or four hundred monosyllables to express ideas, desires, or memories that he felt; and even today peasants need only enough words to express their necessities, which is still more than some Indian tribes, wherein they must light fires at night in order to see each other's faces and hands, since they use gestures and facial expressions to finish their sentences and transmit their ideas.

This is such a natural phenomenon that, on one occasion when I met General Castilla of Peru, he had soiled his frock coat leaning against a window that had just been painted green, which, since he had not been warned, made him vehemently condemn the owner of the house. When he came over to a group of ministers plenipotentiary from the American Congress, he pointed to the stain with his hand, and with his snout drawn in a perfect *moue*, which according to Darwin is a gesture that we share with the monkey, he exclaimed:

"Paint . . . window . . . bad!"

He expressed the exact same thing that has taken me ten lines and seventy words to say; and the truth is that this is plenty to become a popular *caudillo*, like so many that I met in America.

The Bible, with its seventy books, employs just six thousand words, while Shakespeare's dramas make use of twenty thousand.

Languages have developed, then, in the same way as the stars, mankind, and civilization.

How could the law that determines the development of human language be found?

One day, after India had been conquered by the English, a man decided to try and understand the dead language in which the Brahmins' sacred texts are written. He discovered that it was a language related to Greek and Latin, in which there were words common to all three languages—like *pi-tar, mi-tar, pa-ter, ma-ter*—and Jupiter, the supreme god of the Greeks and Romans, was *Dju*, God, and *pi-tar*, father, God the Father.

The family resemblance had been found, but in *Sanskrit*, which is richer and more primitive, the roots and the endings that modify them were different and more visible, such that when analyzing the word *father*, it was found to be composed of *pi*, a verb meaning *to protect*, and *-tor*, as in *protector, creator,* etc. What a lofty notion of the father, the protector of the family, instead of *genitor*, the father according to the flesh! Nevertheless, this word is based on the first action of a child who wants to speak and calls his mother *ma-ma*, and later on calls his father *pa-pa*.

Later on Mr. Calandrelli, author of a dictionary concerning the roots of our language, will speak about this discipline, as well as Dr. Vicente F. López, who has devoted himself to this realm of study.

If the notion of *transformism* in the organic world has not been directly demonstrated in the foregoing sketch—such that over the course of millions of years more perfect plant and animal forms follow those that came before them—being as there is a leap between man and the long and varied line of the quadrumanous animals—amongst whose species all the different parts of the human organism are distributed, at least in their incipient forms, except supreme intelligence and consciousness—there is, nonetheless, a general direction to the succession of the stars, to geological formations, and to the progress of man from prehistory to our own times, as in linguistics, and even in sociology, and in all these different departments of human knowledge, proceeding according the same principle, from simple to compound, from embryonic to complex, from formless form to finished beauty—from all of this the universally accepted theory of EVOLUTION has resulted; and I, gentlemen, adhere to this doctrine of EVOLUTION in these general terms, as a method for the spirit, because I need to rest upon a principle that is both harmonious and beautiful, in order to dispel doubt, which is the torment of the soul.

And now I am approaching the area that I wanted to relate to Darwin's theory, in order to explain the social influence exerted over our times by such shifts in thought.

PART TWO

I feel obliged to beg you a thousand pardons if I allow myself to bring forth certain precedents in order to demonstrate the evolution of thought, the latest expression of which is Darwin.

We, on whichever continent we find ourselves, call ourselves Christian peoples, even though our civilization, art, and law are Greco-Roman. The northern Barbarians also introduced their institutions, perhaps vital organs of the warrior tribe, and furthermore gave us the system of representation that today is so widespread and on its way to being firmly established all across the Christian world.

Since the fall of the Roman Empire, Christianity was the element that brought together semi-educated men, and it was with Christianity that the religious ideals that prevailed in the government of the Hebrew people—to which Jesus belonged—made their way into the Christian government, thus weakening the principles that Greeks, Romans, and Saxons had bequeathed unto us: principles such as the Greeks' freedom of thought and fine arts; the Romans' Senate, Municipality, and law; and the Anglo-Saxons' popular representation in Parliament.

The religious principle dominated everything due to the needs of the times as, following the barbarian invasions, the government was *barbarianized*. The convents saved what ancient books they could, the ones whose parchment was not erased and re-used by the monks to write their Lives of Saints.

Since the kings did not know how to write, the clerics were the only ones who wrote, which is the origin of the word *clerk* in English and *clerc* in French.

The State was religious, and one could say that religion itself was armed with the blade of the law in order to maintain the purity of faith, which was Catholic, so called because of its universality after the separation of the Greek world, which called itself Orthodox.

The kings govern by divine right, and the Pope is held to be the representative of Jesus Christ, who is, in turn, the representative of God.

The great historical event produced by the religious principle is the *Crusades*, in which time after time over the course of two centuries kings, princes, nations, and entire peoples throw themselves headlong against Asia in order to recover the *Holy Sepulcher*, which is to say, nothing, because there was no known tomb of Jesus.

The Holy Sepulcher is religious thought!

Today, reason, as presently formed by the new elements that have come to bear on it, is shocked to think that the satirical poem about the ingenious knight errant and hidalgo, Don Quixote de la Mancha, could be our Iliad of the Middle Ages. Just like him, ten generations of Christians pursued an abundantly and passably absurd chimera.

We will cease to be surprised, however, when we recall that the previous civilization's history was founded on an identical event. In prehistoric times the kings and heroes of Greece traveled to that same Asia to recover beautiful Helen, stolen by Paris, son of Priam, king of Troy, according to the legend.

It is the same event brought about by an identical cause, the fundamental principle behind the society and government. In the Middle Ages, the foundation of the government is Christianity and the divine personage is Jesus Christ. To recover his tomb, then, is to preserve the foundation of the religious government and its tie to the divine.

Greece was founded upon the idea of physical beauty, from which it drew all of its other beauties. Beautiful Helen may be the Hellenic race itself, as the name "Hellas" means Greece. Paris preferred her over Venus, which shows that she was an incarnation of the ideal type of beauty, exactly as it was conceived by the most beautiful race in the world, and that she was also the object of their worship, beneath a luminous sky and a blue sea, on islands and coasts crenulated like the fringe of lacework. In the same way Jesus was the living incarnation of morality and justice sent down from God himself to a sullen people, abused by history since all of the invaders of Asia took them captive, and Alexander and the Roman Reguli humiliated them. Palestine was right next to the Isthmus of Suez, in the path of all the great migrations that stumbled upon it; it was hungry and thirsty for justice, and from out of its viscera emerged a God that was all justice and all love.

With beauty as the foundation of all its aspirations, Greece, despite its low population, produced in less than three centuries the most remarkable civilization of any but our own. The *Iliad* that sang the song of that prehistoric war has had not a single rival in all of the epic literatures.

The statues of Phidias and those that were discovered in Olympia along with half a million works of art that decorated the public squares, palaces, and temples of Greece, Asia Minor, and Rome, have not been reproduced by one hundred million men across twenty centuries in quantity, nor been imitated in their level of perfection, even when the modern artist has the original right before his eyes. Our temples are pale reflections of the Parthenon, and the temple of Ephesus was, without hyperbole, the eighth wonder of the world, given that eight great artists and twelve kings worked together on it.

The modern world left its barbarism behind by merely imitating a classical high relief. The Greeks expressed all forms in the Musics; and they applied the word *music* to writing history, to lyric poetry, to song, to dance, to color, to oratory, to tragedy, to comedy, and to epic poetry. Beauty thus felt in nature, thus expressed in art, infused the soul and was produced in thought.

The Divine Plato was led to God by the belief that there must be a Supreme beauty. For Socrates, justness constituted harmony in human society; and you all know that only the Revolution was able to complete the idea of the simple Athenian who taught us to die for the truth without ostentation.

Pericles, one of their politicians, bequeaths his century unto those twenty that have since contemplated the Parthenon, all that remains of Phidias, and whose name means *administrator* which is what he was historically, just as Helen is the daughter of Greece, Hellas. Their great battles are Marathon and Salamis, battles that halt the course of history and throw the East back across the Ganges. Their oratory is Demosthenes; their medicine gives us a God worshipped down through the centuries, our patron, Hippocrates; Euripides, Sophocles, they all immortalize whatever they touch, and human hands will never retouch their works.

When this plant had to die after the flowering of so many artistic prodigies, it spread its civilization like the aloeswood spreads its seeds, reaching the Orient with Alexander and his war tactics, who with thirty-five thousand men had ten empires made with the shreds that from his tunic were divided up among his generals.

The country that was Greece, up until the Aetolian and Achaean leagues, exhausted into bondage by the Romans, educates its masters and bequeaths unto them the fine arts, the ideal of human grandeur and freedom of thought, inculcating Marcus Aurelius with their stoic philosophy, through which Christianity was born with a more perfect doctrine. From that time on it spends fifteen centuries taming barbarians until the Renaissance, at which point the exclusive reign of religion comes to an end.

Beginning in the year 1400 the Western European world begins to recuperate the Greek elements—forgotten due to the separation of the Orthodox and Catholic Churches—with the books of the ancients that the modern Greeks have saved, and the fine arts begin to be cultivated in Italy, shifting from the Byzantine model of Saint Mark to the Greco-Roman model of Saint Peter.

Let us now examine the state of human knowledge on the eve of the Renaissance.

Teaching is theocratic, for priests alone, given in the cathedral by the schoolmaster canon, by the choir masters in the convents, and given orally for the catechumens in the presbytery.

There is no middle class, there is not a bourgeoisie but rather kings, nobles, bishops, and friars, along with plebs, serfs, and slaves, in the manner of livestock.

To carry out the Crusades, the Church sells admittance to heaven, the kings grant charters to the cities, and the barons give municipal authorities to the towns and villages at the feet of their feudal castles. Thus begins, as a result of the disaster of the Crusades, the modern era, and human freedom is recovered.

I have here a large picture that indicates the progression of thought alongside the dates of historical events.

1330. The Crusades, end of the religious era

A descendant of the Crusades sings Christianity's Iliad, a sublime epic poem about the hallucinations and nightmares of a believer, Dante Alighieri, author of:

The Divine Comedy.

Thus ends the ancient world.

1400. The Renaissance. Gunpowder

In the Battle of Crécy the bodies of men and horses had already been ravaged by the first bombards that, with great thundering, used fire to hurl balls of iron.

Castles and armor cease to protect barons and knights. War will become a plebeian affair, and victory will be won with intelligence.

The nobility is destroyed as a consequence of its uselessness, and democracy arises as a consequence of emancipated labor.

1400. Printing

Gutenberg invents movable type and books are printed in the thousands. No interpretation can be universally accepted, since each person, reading and comparing texts, is his own interpreter.

Emancipation of thought.

Common universal education for all who can read what is written.

The presbyter ceases to teach in the cathedral schools.

Instruction is secularized.

1463. Copernicus

He disrupts and unhinges traditional astronomy, then considered canonical. He places earth among the other planets and reduces the moon to one satellite among so many others that revolve around the seven remaining planets.

The hand of God and the firmament are no longer needed to hold up every sun and planet. *Mathematics and universal attraction come into play.*

The sciences and those who teach them are no longer clergymen.

1495. Vasco da Gama. Columbus. Magellan

They complete the world's geography, verifying that the globe is indeed round.

The theater of human history exits the Mediterranean and enters the Atlantic, the navigation, coasts, archipelagos, and new races of which open up infinite horizons.

Priests lose their preeminence, falling to the rank of boat or army chaplain, or preacher of the Gospel among the savages, but not the directors of the new society, whose explorers and settlers are essentially secular.

Once the rest of the world is known, Brahmanism, Judaism, and idolatry are added to the number of religions.

Antipodes exist, ergo the religious heavens do not.

1493. Alexander VI

One Borgia pope, nephew of another Borgia pope, father of Cesare and Lucrezia Borgia, whom he keeps as a concubine in the Vatican, are the apocalyptic monsters of the worst depravity to ever shame our species.

Because it no longer provides the society with impulses and goals, the moral spirit of Christianity begins to fall apart as kings, princes, and popes give themselves over to shocking states of disorder. In Rome the Messalinas of the ancient empire are replicated, and professional female poisoners proliferate in Italy. That same infidel pope favors in statues, temples, and paintings the resurrection of Greek art, which Raphael and Michelangelo will go on to revive.

One century prior, Dante—the inspired bard of Christian epic—had prophesized, like Isaiah, that God would abandon his people due to the sins of its Shepherds:

È giunta la spada
Col pastorale; e l'una et l'altra insieme
Per viva forza mal conviene che vada.
Di oggi mai che la chiesa de Roma
Per confondere i due reggimenti,
Cade nel fango . . .[6]

1543. Martin Luther

Scandalized by the hideousness of "the Whore," as they call the Rome of the Borgias from that time on, and voicing the complaints of all the people who are pressed and swindled by a system that sells indulgences to pardon all sorts of crimes—indulgences that provided the two hundred million *fuertes* that paid for Saint Peter's—he opens up an era for the examination of the precedents and credentials of those beliefs that permit so much disorder. Without Alexander there would be no Luther.

The Reformation only asks for more Christianity, more morality, more purity, less mysteries, less religious authority and hierarchy.

Historical criticism is born.

1560. Political reaction. Machiavelli

With the restoration of the Communes, with the societies of textile workers in Florence, with the Venetian commerce, with the political freedom stemming from the printing press and public debates, many republics have had a taste of liberty. Machiavelli, a wise and learned man, inspired by the reigning immorality of the period, writes, in the manner of princes and adventurers, on the art of usurping authority and shackling the people. Machiavelli leaves behind a noun: *Machiavellianism*, and many peoples are free regardless.

1565. Religious reaction

The schism that Luther's preaching provoked in the Church, and the secularization brought about by the printing press and the new ways of life that were beginning to open up, encouraged a wounded and retired militia captain to organize an army of astute thinkers and politicians that was to obey *per inde ac cadaver*[7]; with whose aid, says Émile Souvestre, "Captain Loyola intended to stop human progress in its tracks; he opposed the rationalism that was beginning to take hold on the one hand with blind obedience on the other; he opposed the ideas of freedom of scrutiny, discussion, and free government under the rule of law, with absolute monarchy and divine right. To the project that he envisioned, he applied the thinking of a soldier; and he always thought of the Order whose rules he laid out as his army, the army of Christ. This is where the precept of absolute and blind obedience comes from, which is the fundamental principle of Jesuitism."

Jesuitism, as a moral tactic, used its casuistic theologians to proclaim the following:

The ends justify the means.

In the colonization of Paraguay it rehearsed the Medieval theocratic government that intended to restore.

It built on sand. I myself saw orange groves where the Missions once stood.

It has given rise to a word in many languages—*Jesuitism*.

A monumental work of modern literature was also left behind: Pascal's *Provincial Letters*, which constitute the origin of the modern journal of literature and criticism.

The Jesuits still fight to restore the world to its state prior to Copernicus and Columbus, both of whom extended the limits of the sky, the earth, and human intelligence. In our own time Darwin, Agassiz, Gould, and Burmeister continue to extend those limits more and more down into the depths of the earth with geology, and backwards into history with the study of primitive man.

1561. Lord Bacon

He introduced to philosophy the experimental system of induction, as the basis and method of reasoning, abandoning metaphysics, which sought to deduce truth from texts and axioms by means of syllogism. He called this method, with the presentiment of a genius, the *Novum Organum*, sketching out nearly the entire course that modern science would follow.

Theology disappeared from the classroom with the subtle John Scott and the Dominican Aquinas, and the Peripatetic Aristotle.

1561. Galileo Galilei

He measures the oscillations of the pendulum and uses the telescope to observe the sky.

Accused of scientific heresy, he asked the good fathers to tell Pope Urban VIII the lie best to his liking and then stop pestering him, continuing his experiments even at the age of seventy.

His imprisonment, his persecution, and his solemn retraction have enriched human history with one of those vengeful protests that have saved the world:

E pur si muove![8]

and he keeps moving to this very day, just as the sun did not stop to watch some plundering Bedouins fight,[9] as the learned Hebrew scholar Bishop Colenso has demonstrated that this is a mere flaw in translation, to mistake the moon in conjunction for continuing the illumination of the Sun itself,[10] seeming to entail God, at the mercy of each and every adventurer who leads a band of shirtless riffraff, which is what Joshua's men were, historically speaking.

1560. Palissy the potter

If it were possible to see how, inside a human brain, without the patient suspecting it, the ideas that float about formlessly in the atmosphere are settling—like the dust and atoms that we see illuminated in a shaft of sunlight, only to be found later settled on a cornice or in some alcove—it would have been seen in the soul of a potter, painter, glassblower, surveyor, and later maker of porcelain, sculptor, and naturalist, the beginning of the modern age, a commoner being the first one to follow the road laid out by Bacon to reach science by observing everything, collecting everything, testing everything—four hundred different substances to glaze chinaware before discovering lead—and the first to suspect that in all of nature there was a certain order and dependency. Against the opinions of all the learned men to whom he showed fossilized bones, he alone answered that they were the real and true bones of unknown, gigantic animals that had once existed beneath Paris.

Palissy pieced together the first museum from all manner of odd things: minerals, plants, substances, salts, curiosities; and he was the first person to give *public lectures*, gatherings such as this one, with the peculiar twist that he brought together learned men so that they would teach him, or to hear them speak the sorts of nonsense condoned by the persistent schools of alchemy, astrology, and theology.

Artistic and scientific eras

Following this panorama of the fifteenth century, humanity, without its former religious government and civilization, slowly begins to recover the features of Roman law, in its reasoned and harmonious legal codes; with Constitutions, the Anglo-Saxon system of representation; and with the cultivation of Greek fine arts, literature, painting, statuary, and architecture.

Once the peoples' ancient possessions are returned to them, we begin the happy century that we have achieved, *the scientific, constitutional, artistic, free era,* in which, on the American continent, the era of the sciences' application to work comes to fruition, with powerful machinery as our tool, steam and electricity as our motors.

At the end of his great work Darwin puts forth a complement to his theory, which manifests the fecundity of Hellenic civilization's beginnings, and its previously ignored foundation in nature itself: the beauty instinct.

After having attributed the variation among living creatures to the natural selection of those types that are the most vigorous and best adapted to their environment in terms of the fight for survival, Darwin analyzes an even more compelling cause, this being the aspiration toward beauty through sympathy, which has garbed so many animals in such exquisite ways, with ornaments of inimitable elegance and luxury, lit up by all the colors of the rainbow and the metallic lights of smalt.

An Indian girl left her encampment for Río Cuarto, and before presenting herself before society she said in her lingua franca: "making myself pretty first"; and behind a ranch she donned the adornments of the pampas, with her silver plates and her quarter-length pins.

This is the story of the canorous birds and their spotted vestments, of the butterflies and the flowers. The cultivated and elegant woman—since the times of ancient Egypt, India, or Etruria, in whose tombs we find urns with mirrors, pomades, combs, and all the adornments of a lady—is the epilogue to the creation of living things, and is its most beautiful ornament.

Did you know that the gardener's art actually creates new flowers, at its will, by treating plants with lowly flowers to a life of opulence?

A plant's well-being, its civilization we might say, tempers its reproduction, turns certain organs into ornamental accessories, asks the rainbow for its pallet, and little by little alters the plant's forms, centuples its petals, doubles the plant, changes its color, delineates in each profile, undulation, and crook, the line of beauty mastered by Greek art, and there you have ten thousand varieties of rose; the *zinnia*, which from the fields of Argentina has gone on to enrich the gardens of Europe, or the *dahlia*, which has come back from its trip so finely adorned that it hardly recognizes the relatives here in its birthplace, the plains of Mexico and Argentina.

Have I described the story of a flower? No. This is Darwin's evolution and perfection through the desire to *look nice*, which out of the prehistoric squaw made the Venus of Melos, a mere idealization of the living Greek woman model, of that Helen stolen by Paris, of that Phryne who, accused of a great crime undid the brooch of her tunic and showed her form to the Areopagus which, faithful to the Homeric tradition, respected the finest work of all creation and of the Hellenic cult, postponing the examination of her case for one hundred years. What wisdom!

Those who observe the spectacle of an aristocratic dance can confirm whether or not, between the harmonies of the music and the cadence of the movements, the young ladies confirm Darwin's theory, for the betterment and beautification of the race, garbed with all the attractions and seductions of the fine arts, in colors, forms, and accessories. A handful of flowers and feathers, drooping carelessly from one side of the head, causes her to lose her balance and upright posture, prompting her to lift up her face, the undulations of a swan's neck, and the oblique posture that reveals an intelligent life of attention. The peacock's tail must have suggested ideas of majesty to queens and princesses; and so much will a girl do to correct the position of her muscles at rest before the mirror, that in the end she will get what she wants, making herself beautiful, if not herself then her daughters, until she makes hereditary the spruce distinction and the elegance that the Andalusian women brought to this America and which is preserved by our older lineages.[11] Darwinism, pure Darwinism is what common *antiscientific* people call . . .

coquettishness, from *coq*, to act like a rooster, in allusion to his nasty habit of bristling and puffing out his fine plumage.

I have been, ladies and gentlemen, far too kind with me today; but to conclude, I must review the nineteenth century in the same brief terms I sketched out for the fifteenth century, which liberated the human genus with gunpowder and the press.

In this era we are witnessing a period of profound observations and of extensive meditations, as men eagerly give voice to the laws in virtue of which nature, society, and life itself function and exist. Great though not always fruitful efforts are made in order to pry into the secrets of the human brain and, stone by stone, the philosophy of history is being constructed. But the most insightful and profound research is in the field of nature's infinite manifestations, wherein Darwin and his disciples have opened up new and unfamiliar paths.

Perfected methods of observation and experimentation are being applied in the biological sciences, and methods of exact measurement from experimental physics are being applied to physiological research. Research into the theory of spontaneous generation has made an important contribution to the progress of medical science, by discovering that the apparently spontaneous appearance of miniscule organisms is largely due to germs suspended in the atmosphere. This has brought about the discovery of carbolic acid to destroy the countless germs that the air deposits in wounds, and which lead to putrefaction. M. Pasteur continues to explore this new road, opened up to observation.

It has likewise been recognized that many illnesses are caused by the excessive proliferation of microscopic organisms, and one can only hope that the means can be found to destroy these small and terrible enemies, without endangering the patient.

The sciences that deal with prehistoric man have made great progress over the last half-century. No one used to doubt that man had suddenly appeared on the earth six thousand years ago. However, thanks to the Swiss lake dwellings, it has been proven that prior to the period in which man made use of iron, there existed an era in which he only availed himself of bronze, a combination of copper and tin whose having preceded the use of a simple metal such as iron seems improbable. Before he began to use of bronze, man had not discovered any metal to assist his weak hands in the struggle for existence. It has likewise been established that man began to inhabit Europe in the ice age, 21,000 years prior to our own age, and our own scholar, Ameghino, has suggested that this period must have been contemporaneous with the ancient elephant, which would carry man's presence on the earth back to times of incalculable antiquity.

Geologists—among them our learned Mr. Burmeister, as he has shown in his work on *Creation*, which our government has published as an achievement of Argentina's own natural sciences—have become convinced that the phenomena that produced the present structure of the earth are neither the result of periodic violent convulsions, nor some terrestrial energy that determines repeated catastrophes, but rather that the forces that are perpetually at work are powerful enough to produce, over time, such extraordinary results.

When we are faced with the nightmarish grandeur that the mountains press upon us, with their gigantic backs of eternal granite monsters kneeling off in the distance, as one of our poets has said, we think of prodigious conglomerations of atoms, held together by attraction, whose apparent immobility is no more than an equilibrium of effort, and which bring about unceasing transformations that man's brief existence is unable to witness.

There are at least seven hundred thousand species of fossil, of which Lubbock says that three hundred thousand have been described, while Huxley estimated only seventy thousand from the discovery of the *Titanosaurus*—a monster from California's Jurassic layer that measures thirty meters long and nine tall, three times as wide as our streets, and twice the height of our rooms—up to the weak lights that paleontology has managed to shed on primitive life forms.

Geography has nearly completed the rectification of our maps, playing its part in the discovery of the causes that have determined the earth's present configuration. The work of Vasco da Gama, Columbus, Magellan, and Amerigo is finished.

Astronomical science—which with the discovery of the planet Neptune attained one of the greatest triumphs of mathematical genius—has in recent times used spectral analysis to make totally unexpected discoveries about the chemistry of the celestial bodies and their makeup. We now know the metals and gases that fifty different stars have in common with our Sun and Earth, and with the discovery of two hundred and twenty planetoids between Jupiter and Mercury, we can now say that Earth is the two hundred and twenty-third planet.

Our observatory in Cordoba has completed the catalogue of the visible stars that Hipparchus began two thousand years ago. Professor Gould has confirmed—by attempting to measure the light given off by certain stars that do not have stable levels of light, but rather visibly grow and shrink—at any rate, that the constellations are swarms of suns, of worlds in perpetual movement, each and every one of them. Thus sky of fixed stars is a myth, which like so many other myths is but a product of its time.

And now I have at hand what I wanted to tell you all about the evolutionary theory of the elderly Darwin, the youngest of the wise men of the order *Pithecus sapiens*, among the branches of which he has classified man as a mere variety.

In paying homage to Darwin's memory out of the gratitude of this part of humanity, for the good he has bequeathed unto us with his rectifications and discoveries, I believe that we owe an honorable mention to those who, in other fields, have lifted up in this America a part of the veil of the mysterious Isis of scientific truth.

Honor to our compatriot Benjamin Franklin, who *eripuit coelo fulmen sceptrumque tyrannis*,[12] Morse and Edison being mere executors of his will.

Honor and encouragement to Pliny's successor, our learned Burmeister, with the true *History of Creation*.

Honor to our astronomer Gould, who has finished Hipparchus' inventory and restored the movement of the stars, as Copernicus did for the earth.

Honor to Agassiz, who with the illustrious help of a wise Emperor completed his work on fish, the first vertebrates to populate the waters of the abyss out of which birds and reptiles emerged.

Honor to Lubbock, the historian of the life and ideas of America's savages with the *antiquity of man*,[13] which our Ameghino would push back a few centuries further, considering the recent observations from Europe along with his own, until he sinks into the depths of the Miocene layer.

Encouragement and glory to the workers across the whole of our America for helping human science make progress, until down the Mississippi, the Amazon and the Plata, like the triumvirate of the active modern movement, a new American race flows out to the Old Ocean, armed with machines to substitute the organic lack of claws, and brandishing the lightning bolt that it has made its own, until it returns to the Old Land, its mother, in the form of free institutions, the rudiments that Egyptians, Greeks, Romans, and Saxons made for us, and that were brought to us by Puritans and Castilians.

NOTES

1. EN: Possibly refers to the *Cordoba Durchmusterung*.
2. EN: Mr. and Mrs. Agassiz, *A Journey in Brazil* (Boston: Ticknor & Fields, 1863), 42.
3. Ibid., 297.
4. TN: This quote from Huxley's series of lectures, *Six Lectures to Working Men, at the Museum of Practical Geology* (1863) is drawn from two different lectures of the

series, the bulk from "The Past Condition of Organic Nature," and the last paragraph from "The Method by which the Causes of the Present and Past Conditions of Organic Nature Are to Be Discovered. —the Origination of Living Beings." Furthermore, all of the selections have been heavily edited down without any indication to this effect in the text, all ellipses ([. . .]) being my own. However, the only clear instance of manipulation is in the last sentence, which in the original, after a long examination of the facts, reads, "I must content myself with saying that *I do not at all believe* that these are the oldest forms of life" (emphasis mine).

5. TN: This sentence is incomplete in the original as well.

6. TN: "The sword is joined/to the shepherd's crook; and the one with the other together/should by all means be avoided to join./Little time is left before the church of Rome/for confusing the two regiments/falls into the mire" (translation mine).

7. TN: Latin phrase used by Ignacio de Loyola meaning "like a corpse."

8. TN: Italian, "And still he moves!"

9. TN: see Joshua 10:12–13.

10. TN: In other words, the translators of Joshua mistook a description of a lunar eclipse with the sun having stood still in the sky through the night. This passage is obscure in the original Spanish as well.

11. EN: This is essentially Jean-Baptiste Lamarck's (1744–1829) theory of evolution, namely that changes in an organism caused by exertion can be made hereditary. This theory was in fact disproven by Darwin.

12. TN: "He seized the lightning from the sky and the scepter from the tyrants"; said of Benjamin Franklin by Baron de Turgot in 1778.

13. TN: John Lubbock, English banker, politician, biologist, and archeologist, was the author of *Prehistoric Times, as illustrated by ancient remains, and the manners and customs of modern savages* (1865). The *antiquity of man* mentioned here by Sarmiento refers in one way or another to *Geological Evidences of the Antiquity of Man* (1863), by Charles Lyell. It is unclear whether this is a mix-up on the part of Sarmiento, or whether he is alluding to the real controversy between Lyell and Lubbock, who accused the former of plagiarizing his articles in order to compose his *Antiquity of Man*, as it was popularly referred to at the time.

LUZ DEL DÍA'S PILGRIMAGE

(1875)

Juan Bautista Alberdi

VII. QUIXOTANNIA, OR SOCIALIST COLONIZATION IN SOUTH AMERICA

It is good to keep in mind that every European who moves to America becomes freer of spirit, comes to think more of himself, ascribes to himself greater worth, and often goes so far as to become vain and pompous. Don Quixote is no exception to this rule. America has driven him even madder in terms of his characteristic ambition and presumptuousness. His madness has changed its topic, but not its nature. Instead of being Don Quixote of La Mancha, he is Don Quixote of Patagonia; which is to say that the flight of his fancy has known no limits since he has found himself in that favorite land of reckless essays, of fantastic experiments, where all utopias are put to the test, and where even the sanest turn into moderate Don Quixotes.

Settled down in America as never before in Europe, he is the owner of a relatively large ranch, populated, as one usually sees there, by thousands of useful animals of all sorts, horses, sheep, cows, birds, dogs, regular living quarters, without a care for any concerns over his means of living, and living comfortably; since he has had ample time at his disposal, Don Quixote has taken up the most varied of readings. There is no modern book, no social doctrine, or political theory, or scientific discovery, the appearance of which has escaped his ambitious curiosity. Out of all of this a mish-mash has formed in his unstable and fantastical mind, and the natural consequence has been the same as the one to which he already fell victim in Europe. His readings have upset his head and have plunged him into enterprises and projects beside which those from his life in Europe seem sensible. We shall cite, among others, one which, because of its political and social character, has become a matter of public entertainment in America today. Because there is no true social question in America, nor any motive for its existence, everything down to "Mormonism," in the realm of social innovation, is seen as a mere curiosity and let to go about its free and inoffensive development. And so, Don Quixote has been able to manage his enterprise with absolute freedom, until the day that he stumbled onto the ever rugged field of political militancy, and fell into the claws of criminal justice, which is the justice of tigers, as we shall see at the end of this episode.

VII. DARWIN'S THEORY APPLIED TO SOCIAL REGENERATION

Don Quixote had read Darwin's renowned book, about the Origin of Species, that all of them, according to the proofs provided by the earth's natural history, initially arose from four or six forms in which animal and vegetable life had likely made its first appearance on the surface of our planet; that a law which is particular to organic life had multiplied them *ad infinitum*, and that this law was natural selection, or the spontaneous perfectibility of which the species are capable through the accumulation of improvements created by orientation in favor of organic succession; that according to this law of natural and continuous creation, the human species probably had as its origin another less perfect species, for example the monkey.

Over this theory, which has heated up so many heads, Don Quixote's own began to ferment and came up with what we are about to see, translating it into the most curious and unheard of practical experiments.

If man is kin to the monkey, he said to himself, we have all the more reason to believe the sheep to be even closer kin; and in truth this kinship does man more honor, as the monkey is roguish, indecent, useless, thieving, and immoral, whereas the sheep is the religious symbol of docility and goodness: the sheep gives man life, without living off of man, which furthermore confirms that it is the father of man.

According to this hypothesis, confirmed indeed by the many signs provided by man himself gives that he is a refined species of sheep, there is no reason not to believe that all sheep are prone to turning into true men, if only they are educated in man's social mores and practices. If education is second nature, as the saying tells us, Pascal said all the more rightly that nature is the first education. To give a sheep, as its first education, the education of man, is to give unto it human nature after a certain number of years.

Attributing to a few years, the effect that Darwin attributes to thousands of years, in the transformation of animal species, Don Quixote was not content just to say it, but went about putting it into practice as well. An interest less innocent than science led him along in this matter. He had several thousand sheep and just as many horses and cattle on a ranch that started as a kind of toy and, thanks to the peace that his removed location brought him, in a few years it turned into a sort of principality. The ranch was situated between Patagonia and the pampas, somewhat near the sea and closer to the English colony of the Falklands than to Buenos Aires. This solitude left him with his full sovereign freedom to test out all the mad political and social schemes one could imagine. He knew that Patagonia had inspired Darwin's grand idea about the origin of species. He wanted to have the glory of owing to that same desert the first experiment with practical results of this theory, carried out in the service of that grand patriotic notion, of turning, in just four days, a simple ranch into a federal state within the great confederation of La Plata, through an artificial and hasty effort of natural selection, through a sort of *coup d'état* of nature.

Don Quixote had as a butler a young Spaniard[1] with a modest education and a natural good talent, adorned with a greed so boundless and passionate that it made him believe in Don Quixote's delirium even more than Don Quixote himself. Next to a man with Don Quixote's temperament, isolated and removed from all contact with sensible men, seeing the vast lands and the thousands of useful animals that his master possessed, the Spaniard saw in Don Quixote not just a wise man, but a prince, and in his domains, an unrivalled principality, because all around it there was nothing but desert. This simple position was enough for one to do what the Spaniard did, a Sancho without realizing it; and Don Quixote, who did not have much to choose from, made him Secretary and Minister of State, with whom he grew accustomed to discussing and carrying out, as he went about thinking them up and discussing them, all of the plans about which we are going to give a brief report.

IX. CONSTITUTIONAL PLAN FOR A NATION OF SHEEP

Don Quixote gave his ranch, for the time being, the name and rank of "colony"; his animals, "colonists," and his Spaniard, "Secretary General of Quixotannia," which is what the imitator of William Penn named his colony, and he gave himself the title of "Governor" of his Patagonian Pennsylvania. The ranch hands received the title of "intendants," and the colonists were classified into three departments, namely: "homo-sheep," "homo-cows," "homo-horses." The government of the colony was a representative democracy, with a dumb parliament (for the time being), in which each department was to have a number of votes proportional to the size of its population. The "homo-sheep" had an absolute majority, and as they had more than half of the votes, they made the law. Provisionally, and as long as they knew neither how to speak, nor write, nor read, this was to be done for them by the Governor and the Secretary General presiding over the Colonial Council and Parliament. The laws and decrees were to be given and proclaimed in the name of the people of "Quixotannia," proclaimed to be sovereign and free by their founder and liberator. Sovereignty was to be exercised by the people in the form of "plebiscites," expressed by an "ay" or a "nay," in answer to the interrogatory proposals of the Liberator Governor. After two years, the colony would assume the rank of sovereign and free state and a Constitution to this effect would be drawn up definitively. Only then would it enter into relations with the Republic's national government, or with foreign powers whenever need be.

"Well then," asked Don Quixote of his Spaniard one day, "what does my Secretary think of this idea?" (Because genius itself must modestly mistrust its own advice, added Don Quixote to himself.)

"I must admit, Mr. Governor, that I would like to be Secretary of State, even if it were of a state of sheep, but I don't see what benefit we can derive from this project, and I can also see the drawback."

"What is it?" asked Don Quixote in surprise.

"That they're going to laugh at us, and they're going to think we're crazy."

"Who?"

"People from the other nations."

"Why is that?"

"Because of our presumption in forming a political state, with irrational animals."

"You simpleton!" said Don Quixote. "So do you think that those other states are made up of anything but animals? Above all, allow me to point out to you that you are behind in natural history when you speak of 'irrational' animals. They no longer exist, according to modern scholars, who have changed everything. Today it is a known fact that all animals contain a dose of reason and that in this matter it all comes down to a question of quantity; thus, democracy has invaded even the domains of natural history. All that we are going to do is repeat a phenomenon that exists all over America, and even the whole of the democratic world. I'm not the one who says so. The Darwin of modern democracy, Tocqueville, once said these words, which define our democracy in progress, just as they define all nascent democracies: 'an immense multitude of similar or equal men . . . Over them hovers an immense and tutelary power, which takes upon itself only the charge of protecting their possessions . . . This power is absolute . . . It does not tyrannize, but it gets in the way, restrains, stuns, dulls, and, in short, reduces every nation to nothing more than a fold of timid and industrious animals, whose government is its pastor.' And this is not an insult to democracy, as Tocqueville is its own apostle.

"For me," says Don Quixote, "far from an insult, this is its greatest praise, because a democracy of timid animals is a mine heaping with gold and limitless power for the one who governs. The only difference separating the nation of 'Quixotannia' from the other nations whose laughter you fear, is that the inhabitants of ours are citizens in the form of sheep, whereas the others are sheep in the form of citizens. In their political conduct, do not doubt that they are all similar, but with this difference, that ours are the most modest: though they are of more use to civilization, they are not so presumptuous as those who think that they themselves represent it, just because they know how to abuse it at every step."

"But in any case, Master," says the Spaniard, "you can't deny that the other sheep know how to speak, read and write, discuss and vote gathered in assemblies, and use weapons, in short, to more or less exercise their sovereignty well, as if they were men."

"Allow me to give you another word of advice, before responding to you," says Don Quixote. "Don't say 'the other sheep,' because they are so vain that if they find out, they could give us a thrashing. To distinguish them from our own, it will be enough to say 'the others,' or if you like, the 'Saxons,' because all politician sheep are, or believe themselves to be, Saxons of liberal origin.

"All the better if our 'Quixotannian' democrats don't know how to read, write, or speak. This way they better exercise their sovereignty, because they will be obliged to exercise it through our mediation, and we exercise it, as is only natural, first for our own benefit, and then for theirs. In order to delegate it to us and make, to this end, their fundamental laws, they needn't know more than these two words: 'ay' and 'nay,' or if you like, a simple, a mere 'ay.' The 'ay' of the modern nations is the 'fiat' of political 'Genesis': they make the law, just as God made light: with a word. Or if you would like a less stern comparison, the 'ay' of the nations is like the 'I do' of girls, whose being automatically dictated and repeated does not cease to make it the law that governs the freedom of their entire lives as married women. Very well then, as to this 'ay,' that makes families and nations, that creates kings and emperors, that makes constitutions and codes, the people of our 'Quixotannian' nation know how to pronounce it in the most sovereign of ways, and you hear it at every step you take, as soon as they make us out, they set about greeting us affably and respectfully, repeating 'ay,' 'ay,' which in their cold temperament is the equivalent of 'Good!' 'Good!' 'Agreed!' 'Favored!'"

"In that case I've misheard it," says the Spaniard, "since what I thought they were saying was 'mée!' 'mée!' which in French means 'but' [mais], which introduces negation, dissent, protest, resistance, in short, the opposite of ay. Be careful, Master, that you don't open us up to a negative plebiscite on account of a faulty ear!"

Don Quixote laughs at his Secretary's ingenuousness, or rather his linguistic ignorance.

"You have overlooked something quite simple, which is that, as they are of Saxon origin, our sheep (I mean our citizens), speak English, and this word that to you sounds like 'mée,' is, in reality, 'yes,' the English word for '*sí*.' In the absence of their country of origin, our Saxons of 'Quixotannia' have slightly varied the pronunciation. Being of the Saxon race, our nation will at once be free, governable, and civilized. As it descends from 'John Bull,' it will rightfully be 'Juan Lanas,'[2] just as 'King Wool' will be a first cousin to 'King Cotton.'"

"Governable," says the Spaniard, "they sure are, because we only need six dogs to keep twenty thousand sheep under control. But I need my Master to explain to me: just how can a nation of timid and unarmed animals be free, and how can a nation made up of animals be civilized?"

"A good example is the quickest and most convincing professor. England, which is the homeland of liberty, is the homeland of sheep *par excellence*; therefore, the sheep represents liberty, precisely because it is docile and unarmed; which is to say, because it represents peace."

"Pardon me, my lord, if I allow myself to make a respectful observation: isn't the lion the symbol of England, on its coat of arms?"

"Yes, but one might point out that the only lion in all of England is that one on its coat of arms, whereas the land is covered with sheep. India, on the other hand, is the homeland of tigers and lions, which does not keep it from being the slave of the country of sheep.

"The lion is an animal that is armed for war from its birth, and far from being free, for this reason, it rather lives as a slave in cages of iron. It is only free in deserts; where liberty is of no use. The sheep, on the other hand, is never manacled nor encaged, and the sole reason for its freedom is that it has no weapons with which to attack the liberty of others."

"But, I believe, if I'm not mistaken," the Spaniard persists, "that the British Army is not made up of sheep, and that if our only soldiers are our kind and woolly citizens, any enemy would be more than happy to go to war against us, because he would eat good meat without working for it, and have good mattresses to sleep on."

"I do not deny," says Don Quixote, "the need to form a standing army made up of good dogs, while we wait for our citizen sheep to turn into men, by the law of natural selection, helped along by us with top maturatives."

X. DIFFICULTIES OVERCOME

"And how many months, or how many years will it take, according to this great law, for our four-legged citizens to walk, dress, and talk like us?" the Spaniard asks Don Quixote.

"That matter is of no importance," replies Don Quixote, "for the civilization of our nation. Whether a sheep walks on two feet or four hooves; whether he wears cotton, like us, or wears his natural padding of wool; whether he says many words or only speaks one, he will always be a sheep, as long as he maintains a sheep's character. What makes a sheep is not its form; it is its limitless abnegation; and two-thirds of the human species easily exceeds the sheep in this quality. As the people of our nation stand today, they are better prepared for order and progress than the most warring states in South America. There is more achieved for the progress of American liberty by the docility of our sheep than all the courage of our tigers in the form of soldiers.

"I don't know whether or not my admirable Secretary has noticed," continues Don Quixote: "it is a fact that South America's civilization is better represented by animals than by men; by its sheep and its cows, which give it their wools and their hides, in exchange for whose animal products, Europe gives it the products of its manufacturing. The horse, which populates its fields, represents, like steam, the motor force that produces its locomotion and movement. If a material symbol of civilization is its means of communication, it is a fact that the free communication without a set path represented by the horse makes it a vehicle that symbolizes civilization one thousand times more than the 'wagon' of the railroad in both abundance and economy. In South America animals play the part of the slaves in antiquity: neutral in social and political warfare, they provide our sustenance, while we rich and poor spend our lives quarreling over power and wealth. The horse and the ox represent the motor force of industry, in which they are better motors than the arm of man, and nearly as strong as steam itself, since five hundred horses represent the strength of a first-rate steamship. Motor force is gold, bread, wealth, life, civilization. Here the paths are refined. The path of blood is substituted by the path of iron. Yet the horse, which as a locomotive powered by blood settles the pampas and the desert, degenerates instead of improving under these governments that call themselves civilized. We will treat the horse and the sheep as fundamental agents of civilized man; as halves of our own selves; in a word: as our neighbors, under Darwin's law, which is broader than Christ's law in the naturalist's sense, even if it not be so for the theologian."

"I see only one danger in this, Master," says the Spaniard, "and already our cook has brought it to my attention. Since she has accepted the idea that sheep are our natural kin, her conscience has asked her if we can eat our fellow kin without being guilty of the wretched vice of cannibals."

"Just as men eat one another, and even their own friends, in the case of shipwrecks, that is, in cases of dire need, and don't become cannibals because of it," says Don Quixote. "But, we can still maintain the old system for as long as it takes the sheep to adopt our figure and likeness. In this one point, that is, to satisfy our hunger, we will still maintain the notion that they are not men, though they be citizens."

The Spaniard savored this solution, though he couldn't help but observe to himself that his chief's political morality was not of the highest order; and from then on he began to suspect that his ambition was turning him just a bit Machiavellian. Everything will be fine, he said to himself, as long as the precedent isn't applied to me in any extreme case.

"There's another objective that's occurred to me, Master," he said to Don Quixote: "since the death penalty has been abolished by the Constitution, how can we kill a citizen even if it happens to be a sheep?"

"Let's be clear," says Don Quixote: "the death penalty has been abolished for political crimes, but not for private crimes."

"But, as far as this case is concerned," says the Spaniard, "it amounts to the same thing, since what crime could a sheep possibly commit?"

"A property owner does not commit a crime either," replies Don Quixote, "and yet, his property is taken away from him by eminent domain. By the same principle it is lawful to expropriate the life of a sheep for public use, following due compensation of its worth."

"And who will be paid this prior compensation for the life that is taken?" asks the Secretary.

"The sheep itself, naturally," says Don Quixote, "with abundant pastures, as when we fatten up a pig to slaughter and eat it; furthermore, we have one more resort from high politics, in order to save the law's morality, which is to accuse the sheep of some capital crime, such as sedition or treason, in order to justify its necessary death. We will find a good lawyer to see to this department, or, better still than a good lawyer, a good pedant or pedagogue to deliver the sentence with his amusing erudition, and quell the panic of the sheep left behind to await their turn."

NOTES

1. TN: *gallego* ("Galician") in the original; commonly applied metonymically in Argentina to refer to Spanish people generally.
2. TN: The word *lana* means "wool" in Spanish.

HISTORY OF CREATION

(1843)

Hermann Burmeister

LIFE, ITS NATURE, AND ITS QUALITIES

To ask oneself the reason behind the earth's living creatures would be as absurd as investigating why matter exists in its own particular form. Such questions are foreign to the realm of scientific research, whose true aim is to explain what phenomena *are like*, but not to say *wherefore* they obey one law and not another. In all of our research we must not concern ourselves then with anything other than understanding the facts that are presented to us, and attempting to discover the laws that preside over their existence and their evolution; but we should not look for the causes—forever incomprehensible to us—that have produced them in one way and not another. Let us be content, then, to write down that the earth was populated with organisms as soon as it reached a certain level of development, and we shall conclude, given the absolute prototypal[1] correspondence between these organisms and the ones we see today, that in those different periods analogous conditions have prevailed around the globe, and that all living creatures without exception in all periods of time have been subjected to the same primordial idea.

If we examine the idea or concept of the prototype generally, we come to realize that in organic nature, as in inorganic nature, there is a certain concrete form,[2] unique to each object, that serves to outwardly define its individuality or personality. These forms are not arbitrarily conceived phantoms, but rather descend from one or many primitive forms. They are fashioned according to exact mathematical schemata and obey numerical proportions. This character is the same for all the independent and isolated individualities that we call *natural bodies*, though at the same time it is the only morphological property which all of these bodies have in common.

Followed to its final conclusion, we shall recognize the distinction between the forms of natural *inorganic* bodies and those of natural *organic* bodies. The former inorganic ones are mathematical forms, not only in terms of their type or schema, but of their entire constitution as well. They are not delimited, then, by anything but mathematical entities, by *surfaces*, *lines*, and *points*. Natural organic bodies, on the other hand, alongside their mathematical schema, have edges that constitute non-mathematically determined surfaces which are truly unique and, as such, *organic*. The mathematical lines and points are not to be found in them as structural elements. In this way there is, then, a marked opposition between the shape of natural inorganic bodies and that of natural organic bodies. But this is not the only difference that separates them; there are two other still more distinctive features as well, of which the first is *material* and the other is *functional* or *ideal*.

Natural inorganic bodies are *materially* distinct in that their elements are directly assembled according to the concrete form that corresponds to the particular body to which they belong. This phenomenon is referred to as *crystallization*, and the shapes of inorganic bodies, when abstracted from their material composition, *crystals*. Natural organic bodies, on the other hand, only incorporate the substances of which they are composed into their mass within tiny independent vesicles, called *cells*, and they transform all organic matter into cells before it comes to form a direct part of the *tissues* that make up its structure. Organic bodies, as far as material differences are concerned, are entirely composed of isolated *atoms* or particles, and as such have an *atomic* structure; inorganic bodies never have such a structure and are always *homogeneous*.

The *ideal* distinction between the two groups of natural bodies lies in the way in which each of them maintains its own unique form. All inorganic bodies can only maintain themselves through the complete immobility of their form and composition and, to further their existence, must necessarily remain utterly stationary in their original state. Their existence as natural bodies, their duration in time, is made possible only through their immobility; it is destroyed by those modifications to their form and composition that carry with them the death of the individual. Organic beings act altogether to the contrary. Their preservation lies in the continuous wear on their basic parts and the absorption of new matter to compensate for that which has been used up. As a consequence of this phenomenon they continually modify their form and substance through the expansion of certain limits. They shrink and grow in volume, in weight, in dimensions, and in composition, and they move as free and independent bodies within their sphere of existence. Thus, due to this change, which is in fact a true movement, they can be defined as *variable natural bodies*, or even be called, in a more characteristic fashion, *periodical* (following Fischer and Link)[3] or *cyclical*[4] bodies, because their changes are subject to phases that reappear in succession, i.e. periods. A constant periodicity dominates the existence of all organic bodies, and its manifestation comprises all that we understand the words *life* and *vitality* to mean as far as they are concerned. *Death* is nothing more than the final act; it is a necessary consequence.

[. . .]

To these facts we will connect important and extremely interesting considerations regarding the primitive origin of organic and inorganic bodies, considerations deduced from the earth's elemental substances and the differences that proceed from them. We have characterized inorganic beings according to this point of view, saying that they took their own particular form as soon as material elements came into being, and no external obstacle, such as the absence of a required temperature or a necessary calm got in the way of their formation. [. . .] With living organisms the story is completely different, at least in the present era. Their birth does not at all depend solely upon the simple mixing of their basic parts; rather the cause is always some other influence that has not yet been discovered and, as such, about which we can say nothing. It would appear that this influence can only be enacted by another similar living organism; and, what is more, it depends not upon its will but rather obeys eternal and unchangeable laws. We know all of the substances that make up a homogeneous organic material such as *albumen*; but we cannot artificially produce albumen. On the other hand, nothing could be easier than to create true water, mixing oxygen and hydrogen in the appropriate proportions and passing electrical sparks through this mixture. And yet all hens effortlessly produce the albumen that surrounds their eggs' yolks. In reality, the phenomenon is independent of her will and is only brought about when the yolk penetrates the oviduct. As the state of our knowledge currently stands, we cannot comprehend the birth of an organic material with the help of inorganic elements without the influence of a pre-existing living organism. We find ourselves, as such, immensely ignorant as to the initial origin of beings on the surface of the earth. It has been said that new, independent, and completely different organisms could spring from a living organism, formed out of a simple organic substance, without having to pass through the intermediate germ and egg. But this theory is far from having been explained and does little to further the question, given that a living organism is still needed in order to carry out a special action upon the matter. The formation of new and strange organisms inside other types of living beings without a germ or an egg carries the name of spontaneous generation (*generatio originaria seu equivoca*). This is the opposite of *sexual reproduction* by means of a germ and an egg, which is the normal means of generation among organic beings, from which it can be deduced that the former is an impossibility. But this law has been given far too wide a scope, whatever the supporters of *equivocal generation* may have said about it. Based on this birth of one organism from a different one, this group deduces the possibility that all organisms were generated via the same course in primordial times.

However, for the present era they only allow for the ability to produce inferior and imperfect organic bodies with the help of elemental bodies. Whether or not this system is founded upon positive bases is still problematic, though the majority of our contemporaries' voices have spoken against it. Nevertheless, we grant it some value because in reality there is no truly scientific objection to it, and without it the origin of the earth's organisms cannot be explained other than through the immediate intervention of some higher power. But the introduction of this external intervention, in all other parts of the Earth's development, proves to be totally unmotivated, and even contradicts other scientific results. If we were to allow for its existence when the first organisms arose, then we would have to allow for it to reappear later on in every one of the planet's upheavals; which is clearly contrary to the great plan that the world's order has followed.

Even if *spontaneous generation* as a natural law is a necessary postulate for positive science, we still cannot deny that the most recent experiments show that its existence in the present period is improbable. For a long time they based their arguments, with good reason, on the *entozoans* that live inside the closed cavities of animal bodies and remain inside them for long periods of time. They have no reproductive organs and can produce no offspring. This was believed to be the rigorous proof of their having been born through spontaneous generation; because an animal that has no sexual organs can neither produce eggs nor give birth to young.

However, it was later seen that these supposed entozoans were not born in that spot, but rather were introduced there from the outside with a completely different form, and penetrated their way through completely sealed off casings and into the place where they were to live. It is necessary for the animal in which they live at the outset to die, under any sort of circumstance, in order for these worms to reach their final home. They are devoured inside of their first host by a second one, in whose body they reach their final stage of development and turn into creatures equipped with large reproductive organs. [. . .]

Even though *generatio originaria* has lost the mainstays of its support in the present era, the question of the initial origin of the earth's organisms is not thus torn asunder. The explanation as to why no new animals are directly born at present is simple, once we note that *all of them* possess a sexual apparatus or its equivalent. But what we cannot comprehend is how animals were ever born without some direct act from outside during the long period in which there were still no animals equipped with sexual organs. In the present period, in which everywhere there have long-since lived creatures capable of reproducing, it is no longer necessary for other new ones to spring forth from the bosom of primordial matter. It may also be the case that the material elements from which they could be formed no longer exist because the majority of the organic substance is mixed into living organisms, and no longer able to provide the materials necessary for the birth of new individuals by any means other than sexual generation. In short, the struggle in which all organisms are engaged in order to obtain nourishment bars any accumulation of free organic matter, since even dead organisms themselves serve as food for a great number of living organisms, and since, as far as one can tell, only a very small portion of their mass returns to its inorganic elements.

Now we find ourselves faced with the following question: from whence does the organic matter that gave birth to living creatures come? There is a second related question, about the way this material was formed.

It is not difficult to answer from whence primordial organic matter came, since we know that in the era in which organisms were born the earth already had a mixed atmosphere of oxygen and nitrogen, was covered by seas whose temperature by this point must have dropped to 60° Réaumur, and which contained, just as with the atmosphere, large quantities of carbonic acid.

The admission of this final point is an indispensable condition in order to explain the carbon that is contained in all organisms and which evidently comes from this carbonic acid. The observation we made earlier regarding emanations of free carbonic acid in volcanic regions and eruptions shed much light on the origin of this body, which probably came out of the earth's interior abysses during and after the eruptions first of plutonic matter and then of it itself. The conditions necessary for the formation of organic matter were, then, present and incredibly rich; nitrogenous compounds, such as nitric acid and ammonia, filled the atmosphere in high proportions, since organisms had not yet been able to fix them in their tissues.[5] The existence of free hydrogen in the period in which living creatures were formed is not admissible following what we know about its affinity for oxygen and, moreover, it is not necessary given that in all vegetable substances there is never a quantity larger than that which is necessary in order to form water with oxygen. It would, then, have been enough for water to be simply absorbed in order for hydrogen to get to plants and from them to animals. The different solid substances that exist in living organisms were surely dissolved in the water, especially in the case of calcium carbonate and silica, the first of which has a special role in animals, and the second in plants. Its presence in the most primitive waters is, according to what we said before, as certain as that of sulfuric and phosphoric acid, of table salt, and of numerous other inorganic bodies that take part in the composition of inorganic substances, or are found in living creatures. When we stop to think that along with this abundance of materials, the temperature could still have reached 60° Réaumur at the moment in which the first organisms were born (this high heat being, in fact, the maximum that living creatures can withstand), and at the same time that moisture was nowhere lacking, we have here a real cause for fertility in the soil and a powerful instrument to imagine the first organic matter that was made with the help of all these varied elements. Let us suppose then that the first organisms were born around this period of cooling.

The means by which they were formed is, in all other respects, an enigma which will always remain unresolved and about which we can say nothing precise.[6] It is plain to see that, in this case as well, the most realistic opinion is that which is connected to the workings of present-day phenomena and rejects any intervention by mysterious powers. If we accept, then, that the first creatures were not born immediately with perfected forms, but rather that they appeared, in accordance with the ordinary manner, as individual imperfect young through a series of phenomena analogous to their current means of development, we will have said all that can be reasonably said about their origin, as we are unable to penetrate any further into the details of their birth. Let us be honest—our positive knowledge is insufficient to sketch any strictly permissible picture of primordial organic creation, and the painter who wishes to attempt to fill out the edges has before him a vast field open to his imagination's fantasies. Some may admire the product of such speculations, an entire nation can unite behind a myth that it has invented or received from the outside; but such attempts before the enlightened eyes of the scholar will never be worth more than what they are, which is to say, the nebulous shapes of a delirium, totally hollow and inconsistent, which attempt in vain to reach a true shape though the continual modification of their edges. Ancient day of life's first appearance, whatever your nature may have been, we now have no eye whose gaze might reach you, no sense with which to comprehend you, and *as such* no pen with which to describe you.

NOTES

1. OAN: In the natural sciences, *prototype* is understood to mean the ideal form behind every particular (*concrete*) creature; it does not exist alone and is nothing more than a pure concept. The *avian prototype* is the idea according to which all *birds* are structured; but a *sparrow* is a concrete reality, possessing particular properties that distinguish it from the rest of the birds.

2. OAN: The form of any body is not only the essential condition of its existence insomuch as it can only become a *living reality* by adopting a determined form, but also insofar as it is the only truly *passing* and *perishable* thing about it. When a natural body ceases to be, when a plant or an animal dies, they only disappear as individuals. Their matter, the substance out of which they were made, returns to its erstwhile amorphous form, it dissolves into the water or in the air, or gets lost in the ground as a solid element. Thus matter *does not die*, does not disappear; it is indestructible and eternal; it has always existed and knows no limits of time. The most it can be individualized is by adopting a determined form, but it existed prior to that form, though often in a different manner and in other combinations. So, to conclude, any apparent destruction can only be either a new change of form or else a destruction of substance, which is *absolutely* impossible. The earth never loses any of its material form; it only changes the form in which it previously existed and becomes a new object, understood only in terms of this point of view.

3. OAN: *Propylaea of the Sciences of Nature*, vol I, p. 142. Berlin, 1836. In octavo.

4. OAN: See my *Handbuch der Naturgeschichte* (*Handbook of Natural History*). 1st part, p. 13. Berlin, 1836. In octavo.

5. OAN: It is very interesting to note that elemental inorganic matter never comes to form part of an organism by any means other than breaking off from a chemical combination in which form it has entered, and never by penetrating into the organism directly. The direct absorption of oxygen in animals' respiration is merely apparent, given that the same quantity of this gas is expelled in the state of a compound with carbon and hydrogen in the form of carbonic acid and water, and as such is not used at all by the animal's makeup. Thus nitrogen, like hydrogen, could not enter directly into the first living matter; it had to break off from nitrogenous compounds.

6. EN: In the first note to page 191 (Spanish edition of Juan Ulled, 1900) of Volume I of his *History of Creation*, in speaking about geological formations, Burmeister writes the following: "Further along (v. II, c. II) we will deal with the origin of organisms' general beginnings. For the moment, it will be enough for us

to point out that there are two opposing theories used to explain the differences specific to each of the earth's eras. In the first doctrine, the living creatures of the first periods, which disappeared in the periods that followed, perished at the end of these periods and new species, adequate for the new environment of which they were a consequence, immediately replaced the preceding ones. The second system would only have the forms absent in the following periods disappear. That is to say, that they are not represented in them by analogous species. On the contrary, the analogous species, with their partial modifications and living into the most recent periods, are the descendants of the old ones, whose traits have been modified at the hands of a changed environment. As I see it, the first of these opinions is just as acceptable as the second, because neither of them can present any positive fact taken from historical times in its support. They do not have, then, more than a dogmatic or hypothetical value. Recently, Darwin has taken up the latter hypothesis again with great developments, and following his attempt many of its points have been received with great applause. In spite of all this, we must confess that we can grant no demonstrative power to the arguments put forth by Darwin and his supporters, and that perhaps it would be best to leave this question aside as a matter inaccessible to experience. The hypothesis of transformation is unnecessary to explain the initial origin of all organisms, given that what can be born once can by the same logic be born again under similar conditions. How did this phenomenon take place? We do not know, nor shall we ever know. As such it seems to me to be of no use to empirical science to imagine hypothetical conceptions regarding this problem and get lost in endless controversies about their probability. As for the rest, either doctrine could outlive the other. It has indeed been proven that living creatures undergo gradual transformations, but what I also find to be very true is that all of the differences found up till now between them cannot be explained by simple transformations."

REMEMBRANCE IN HONOR OF DARWIN

Transformism as an Exact Science Lecture

(1889)

Florentino Ameghino

Gentlemen:

The conclusions at which I arrived in my quick dissertation on the stone age, as you have seen, are openly transformist or Darwinist, however you wish to call them. This first appearance of the art of breaking stones in the incredibly remote past, and this continual, slow development of stone industry across the geological eras, is the theory of evolution, of which Darwin was, in our time, the most skillful and commanding defender. I stand before you, gentlemen—without having intended to, and forced by the circumstances—in the midst of Darwinist territory . . . and the Master has just paid his tribute to nature, which is what such an act is according to the law of Malthus. The telegraphic wire has not yet finished relaying to us the final mournful echoes of the posthumous honors paid to him all across Europe . . . How could I pass by without pausing before this commanding intellectual beacon? No; I could not do so. I owe him a remembrance in the name of the Argentine Geographic Society, which has done me the honor of inviting me to give this lecture, and likewise I myself owe him a remembrance, because I was one of the first disciples from the Argentine Republic to adopt the ideas of the renowned Master . . . In fact: my ideas in this regard have been known by my friends for years; since I was in school, I could almost say; and in this vein I might recall for you of a curious anecdote, one that is little-known and that I myself had almost forgotten.

Some eight or ten years ago, if I remember correctly, my friends thought that my transformist obsessions were absolutely ridiculous—they could hardly believe my assertion that there was a Darwin and a Huxley who publicly defended such ideas, and so attributed them to me alone. Determined to keep me off of the road to hell, they decided to make a fool of me in order to do so. At that time there was a satirical/burlesque paper in circulation called *El Cencerro*,[1] of which only a few issues were printed.

One day I received a copy in an envelope: in it there was a large number of words arranged so as to form a maze with the following title above it: "Lesson on Modern Zoology by Professor Ameghino." Days later a friend, who swore he had not been the author of the little jest, which I did not really mind, gave me the key to reading it. I do not remember its literal content, but in effect it said the following: *Men before the Deluge were quadrupeds and only afterward did they become bipeds*. These ideas, which in order to ridicule me they printed with vulgar and, one might say, foul words are, ultimately, the right ones, and the same ideas that I currently profess; but at that time I was far from believing that one day I could show them my contingent of material proofs.

Before continuing down this path, I must make a protest against the mass of clamoring anti-transformists who, in their eagerness to fight the new theory and impede it from gaining converts, distribute absurd falsities, such as one that is common here, that Darwinists would have man descend from the monkey: the brachycephalic Asians from the orangutang; the dolicephalic blacks from the gorilla; the central African pygmies from the chimpanzee. Such attempted perversions, I will call them, are absurd. Furthermore, I might add that it is not at all gentlemanly to attribute nonsensical remarks to one who has not uttered them. Neither Darwin nor his predecessor Lamarck, nor his disciples Huxley and Haeckel, nor any naturalist of the transformist school, has said that any of the current human races descends from any of the current species of monkey. What transformists argue is that of all creatures in general, and each species in particular, none has sprung up just because, all of a sudden, overnight; that nothing is created out of nothing; and that as a consequence all creatures must have ancestors. In regards to the superior animal forms in particular, of which we are the apex, what this school upholds is that man descended from an inferior extinct form, that the current anthropomorphic monkeys descended from another, also extinct form, which in turn undoubtedly had its origin in a primitive ancestor from which the early precursors to man diverged in some extremely remote era.

Now that I have promised to tell you something about Darwin and the theory of evolution I find myself in a quandary, because the theory takes in the entire Universe. It is too much for one man and for many. As such, I will limit myself. I will not speak to you about evolutionary theory as it relates to the formation of the stars, of languages, of religions, or of nations; less still about the transformations of plants. It is too vast a field, and I am neither an astronomer, nor a linguist, nor a botanist, much less a philosopher or a politician. I am an anthropologist, and, above all, a paleontologist. As such I will deal with transformism as it pertains to my own domains, without repeating well-known facts, but rather presenting to you in a few words some of the new materials that prove Darwin's theory to be a fact, and even allow us to place it among the ranks of such exact sciences as astronomy, given that the facts and phenomena with which both are concerned can be reduced to formulas and laws. These laws involve a level of precision that allows scientists in both fields to predict findings and discoveries from their desks, availing themselves of nothing more than numbers . . . Do not laugh, gentlemen, at what appears to be nonsense; give me your benevolent attention for just a moment longer and then you may judge for yourselves. This is the best homage that you and I can pay to Darwin's memory.

You all must know that Darwin can be considered one of our own thinkers, as the discovery of his theory is linked to the history of our own scientific progress, since it was here, in our midst, where he collected the materials for it and had his first ideas. And, by what is truly an extraordinary coincidence, it is here, only here on the pampas, where its clearest proof can be found, due to reasons that are within everyone's reach.

One of the foremost objections made against Darwin's theory is that the many intermediate forms that should link current forms with extinct ones are missing. Very few such intermediate forms have been found up till now on the old continent; and few will be found, because the geological formations there have been displaced in all directions and partially destroyed, such that they are not found in a continuous series. What we have, in a sense, is a book in octavo, four fifths of whose pages have been torn out: the history, long ago written there, can no longer be read. Things are different on the pampas; at one time it was believed that the study of geological formations was more difficult here than in Europe, and this was a mistake. What is for certain is that the root causes of the great geological movements were more powerful and uniform here and that, consequently, their effects are presented to us in vaster proportions and in a less-interrupted series. What we have here is an enormous book in folio, out of which pages have only been torn here and there: the history written within can be read almost straight through. Add to this that the nature of the pampas' terrain allows for better preservation of organic remains than most European formations, and so it is easy to understand why it is here, where it was born, that Darwin's theory should find its most splendid corroboration.

Time is passing and I must take advantage of it. I will, then, engage my subject directly, taking as the basis of my explanation one mammalian order in particular: Edentata. Today this order is only represented in Argentina by a short list of small-sized animals of the armadillo family, known by their common names: *mulita, mataco, peludo,* etc. However, in past eras the pampas were inhabited by an immense number of edentates of all shapes and sizes. The exuberance of forms and ways of life exhibited by these animals at this time was both prodigious and astonishing. When studying this particular type of fauna it seems as though one were witnessing the appearance, development, and extinction of an entire mammalian order, Edentata. This is one of the richest orders in terms of distinct forms, and more differences are to be found within the order itself than between two other genera taken from two different orders. There are, for example, more differences between the Megatherium and the *peludo*, which are classified in the same order, than between the dog and the family Lemuridae, which are classified in two different ones.

The Megatherium is a colossal, elephant-sized animal, four times more robust than the elephant, with molars that form an enormous quadrangular prism, and protected by a carapace. The distance between the two animals is vast, and yet in just a moment you will see it disappear.

In the area of the pampas, we find the remains of an animal that is somewhat similar to the present-day *peludo*: the *glyptodonts*, which comprise five or six different genera, with more than twenty different species. The glyptodonts differ from the armadillos particularly in terms of their gigantic size, in some cases as large as an elephant; in terms of their fixed dorsal carapace, that is to say: without the mobile rings or bands that the present-day armadillo exhibits; in terms of their spinal column, whose parts are partially fused to form a sort of tube; in terms of the shape of their head and, above all, their lower mandible, whose ascending ramus forms an acute angle with the horizontal ramus; in terms of the number of its molars, always numbering eight on either side of each mandible; and in terms of the shape of these same molars, which appear to be made up of three triangular prisms connected at their angles. Some have believed that present-day armadillos are the degenerated descendants of the ancient glyptodonts, and this is also a mistake. In fact, in the same period there were true armadillos as well, of the same shape and size as those of the present-day, though some extinct forms allow us to cross over from these to the glyptodonts. There is a fossil animal, the *Eutatus*, that is very similar to the *peludo*, though much larger in size; and another, the *Propraopus*, that is very similar to the *mulita*, though likewise larger in size. And another extinct genus, the *Chlamydotherium*, constitutes a true transition between the former animals and the glyptodonts.

The *Chlamydotherium*, according to our specimens, was as large as the small glyptodonts or the large armadillos. Its carapace was similar to that of the latter in that it already had some mobile rings around its mid-section. Its mandible is an intermediate form between that of the glyptodonts and that of the armadillos; its teeth are likewise intermediate in terms of shape, and in that the glyptodonts' constant number eight is modified in that it has, like many armadillos, nine in the lower mandible. In this way we can cross over from the armadillos to the glyptodonts without taking any great leaps. Now the task is to link the glyptodonts, animals protected by a thick, solid carapace, to the Megatheriidae, in whose family we find the *Megatherium*, an animal without a carapace. This undoubtedly looks more difficult. However, it is not.

Some twelve or fourteen years ago the learned Dr. Burmeister—who is opposed to the theory of evolution and who, nonetheless, because of his *History of Creation* should be placed among Darwin's predecessors—made a discovery that was of the utmost importance for transformism. He found, near the banks of the Salado River, part of a *Mylodon* skeleton, from the same family as and very similar to the Megatherium. From on top of this skeleton he gathered a large number of small formless bones, like round little pebbles, and with the sagacity of an experienced naturalist he instantly realized that in another time they had been embedded in the animal's skin, forming a sort of rudimentary carapace. We find ourselves, then, in the presence of an animal that is very similar to the *Megatherium* and has the building blocks of a carapace. Now there is only one step left to get from the perfect carapace of the glyptodonts to the rudimentary armor of the *Mylodon*, and I will guide you to it.

The true Glyptodont's carapace is composed of a large number of pentagonal or hexagonal plates joined together by fixed sutures whose outer surface is adorned with shapes or designs. Yet, there is one genus: *Euryurus*, in which the outer surface of the plates is rugose, without shapes or adornments of any kind, and which are not joined together as well as in the previous genus. Another genus, *Doedicurus*, has a carapace made up of completely smooth plates, without any adornments and with large holes that run through from one to the next. These plates were embedded in the skin like the *Mylodon*'s little bones, but with the difference that these were immobile. What remains now is a small gap. An intermediate form between the *Mylodon* and the *Doedicurus*. By induction I had predicted its existence years ago and had provisionally named it *Myloglyptodon*. I will return to this point. Today this animal is known by the name *Thoracophorus*. It has a carapace composed of small symmetrical bones like the plates in the glyptodonts' carapaces, but rather than being joined together by fixed sutures, they are simply placed side-by-side like the *Mylodon*'s little asymmetrical bones.

Now you can see, gentlemen, that I was right when I told you that the distance separating the *Megatherium* from the *peludo* is not an abyss at all, since we can cross over from the present-day armadillos to the fossil species, from there to *Eutatus* and to *Propraopus*, from there to *Clamidotherium*, from *Clamidotherium* to *Hoplophorus*, to *Panochtus* and to *Glyptodon*, with its fixed and solid carapace, and crossing over from there to the *Euryurus* with its unadorned carapace, the *Doedicurus* whose carapace was embedded in its skin, to the *Thoracophorus* with its entirely flexible carapace, down to the *Mylodon* with its rudimentary carapace, and from here we cross over to the *Coelodon*, an animal likewise protected by a rudimentary carapace and which is even closer to the *Megatherium* than the *Mylodon*.

And that is not all. I am going to pause for a moment at the Megatheriidae family, and show you all something even more surprising. Some have been critical of my classifications, saying that I make up an excessive number of species, and that the majority of the forms that I consider as such are mere varieties. Congratulations: I accept the criticism, because I do not care whether these forms are called species, races, or varieties, or what have you, as it all only goes to prove what Darwin already said: that classifications are artificial, not natural. My task is to distinguish the forms with a name so as not to mix them up, to stake them out and move on to new ones. And of course I can assure you that displaying the extinct edentate forms that I possess from the Megatheriiade family in this way, it is not my species that disappear, but rather the same species traditionally admitted by previous authors and those very genera that were believed to be absolutely different from one another. Follow me for a moment through this quick exposition and you will see. To simplify the comparison I will take into account only a part of the skeleton: the masticatory apparatus.

Nearly all of the genera of the Megatheriidae family have five upper and four lower molars. The two extreme types are: the *Megalochnus* and the *Megatherium*. In the *Megalochnus* the two front teeth are located at the front of the mouth, one next to the other, like a rodent's incisors: they are true incisors separated from the rest of the molars by a long space. In the *Megatherium* all of the molars have more or less the same quadrangular prism shape, with two transversal humps on the surface of the crown. The palate is very narrow and the section outside the molars protrudes forward. I am still unaware of the shape of the molars in the smaller species of *Megatherium*, but in the *Coelodon* they are slightly modified and the palate is not as narrow and elongated. In another unclassified animal that is currently at the Exhibition, the shape of the molars tends toward those of the genus *Scelidotherium*, whose palate is so narrow and protrudes forward so much that it is rather similar to the *Megatherium*. Beginning with the *Scelidotherium leptocephalum*, the molars, though always used horizontally, are modified, and as we move along to the *Scelidotherium Capellinii*, and the *Scelidotherium tarijense*, it can almost be confused with another animal, so intermediate that Bravard placed it in the genus *Scelidotherium*, calling it *Scelidotherium ankylosopun*, and Owen in the genus *Mylodon*, assigning it the name *Mylodon Darwini*. Professor Reinhardt has just given it the new name *Grypotherium*. In this animal, the first molar, although always used horizontally, is moved over so slightly as to hardly distinguish it from the rest. From here we move on to the *Mylodon Wieneri*, to the *robustus*, etc., until we find the *Mylodon intermedius*; in this species the first molar is used as a slight angle, taking on the shape of a canine and forming a transition to the genus *Pseudolestodon*, of which the species *Pseudolestodon debilis* is a close relative. In the *Pseudolestodon* this development of the first molar toward a canine becomes more and more evident in six or seven species leading up to the genus *Lestodon*, of which the species called *Lestodon Bravardi* is the one most closely related to the genus *Pseudolestodon*. In the *Lestodon Bravardi* the first molar is a true canine separated from the other teeth by a long space; and the front part of the palate is much wider than the rear. These traits become more and more accentuated as we move from *Lestodon Gaudryi*, *armatus*, and *Bocagei* to *Lestodon trigonidens*. In this last species the fangs are found in the front part of the mouth and are very large, and the lower set is positioned obliquely. Only one small leap is needed now to get to the *Megolochnus* with its true incisors, which we are spared by crossing over *Megalonyx*, in which it is unclear whether the front two teeth should be considered canines or incisors.

By means of this hardly noticeable transition, the soundest genera, such as *Lestodon*, *Megalonix*, *Mylodon*, *Scelidotherium*, *Megatherium*, etc., would be reduced to mere species, mere varieties. This is more than sufficient to prove once again that our current classifications are not natural but artificial and that, consequently, transformism is a reality.

I am going to move on to another field, to show you the genealogy of some of the animals currently living on the pampas; for example: the common fox, the viscacha, and the guanaco. In the deepest layers of the Pampean deposit we find a fox, which Bravard called *Canis vulpinus*; it is smaller than the present-day fox and its teeth are closer together. At a slightly higher level we find what Lund called *Canis protalopex*; it is a modified form of the prior fox. This modification continues in remains from successively higher levels until we come, through imperceptible gradations, to the present-day field fox, *Canis Azarae*.

In the same deep layers of the Pampean deposit we find a viscacha that is considerably different from the one alive today: Burmeister has called it *Lagostomus angustidens*. As with the ancient fox, the ancient viscacha is considerably smaller than the present-day one, and its sagittal crest is not as raised. At a slightly higher level we find somewhat modified remains; the animal is more robust and the sagittal crest is wider and higher. In the upper Pampean layer still larger specimens are found, with a higher crest and wider incisors. In the post-Pampean deposits it hardly differs from the current viscacha. In this way we can move in succession from Burmeister's *Lagostomus angustidens* to the present-day *Lagostomus trichodactylus*, and you can be certain that the latter descended from the former.

Just the opposite occurs with the guanaco. The animal that represents it in the lower strata of the Pampean deposit is three or four times larger. The present-day guanaco has four molars in a row on either side of its lower mandible. The ancient guanaco, known as *Paleolama*, had five. In remains from successively higher levels the size diminishes; and along with this the size of the supplementary fifth molar. In the post-Pampean deposits, that is to say, in the relatively recent period, it has nearly the same size as the present-day guanaco, but the supplementary molar, though small, is present in most cases. In the present-day guanaco it has disappeared, but when the animal is very young this trait reappears and we find the small molar that was exhibited by its ancestors. I could extend these observations to other animals from our pampas, but the examples I have cited are sufficient and I am going to move on to an even more conclusive class of evidence that makes me consider transformism to be an exact science, one that will eventually be able to solve everything by means of equations, multiplication and division.

At one time it was believed that the animal kingdom was arranged in a continuous linear series, like links on an immense chain; this was called the zoological scale.

Darwin and his disciples called it the animal series; and they compared it not with a chain, but with a vast tree, vastly branching out, and whose branches diverge from one another as we approach the present day, and converge in a common trunk as we enter the depths of the past. As for me, I too would compare the animal series with a tree, but with the sole intention of finding new laws to corroborate the theory of transformism.

The tree's trunk would represent the first imperfect creature or creatures that appeared on the planet. As the tree develops, the trunk branches out and the struggle for existence inevitably begins between the different branches that fight over air, light, heat, and moisture. The branches continue to grow and the struggle broadens, but not all branches share the same fate. The secondary, tertiary, quaternary branches, and so on, represent the classes, orders, families, etc. The final ramifications represent the species; and the leaves that are periodically renewed are the individuals. In the struggle for life it often happens that the first branches, deprived of light by the others, cease to develop; these represent ancient inferior species that have continued unchanged up to our times. Other branches, singed by lightning or broken by a hurricane (both of which are, in this case, the equivalents of natural disasters), or due to any other cause, dry out and their remains fall around the base of the tree; these dry branches represent animals that are now extinct. One day we pass by the trunk of the tree and gather the dry branches that we find on the ground, in order to build a fire or what have you; this debris represents the fossil remains that we find buried in the depths of the ground. The final ramifications of the tree, the ones that continue to develop normally, are the species that currently exist. From all of this that I have explained in this comparison, two laws of utmost transcendence for transformism or Darwinism can be deduced. First: *That many species and genera of animals have disappeared not due to transformation, but rather extinction, without leaving behind any descendants*; these are the tree's dry branches. Second: *That all the present animals must have predecessors in past geological eras.* I am often asked about the modified descendants of *Typotherium*, of *Toxodon*, and of *Megatherium*, or of *Glyptodon*. These are dry branches of the tree, gentlemen, which have gone extinct without leaving behind a line of descent. I can tell you that the second law has been completely proven in the present-day province of Buenos Aires. All of the current mammal species of the pampas, with very rare exceptions, have been found in fossil state in the same spots inhabited by their present-day descendants. The scientific works of Dr. Burmeister, of D'Orbigny, and of Bravard have shown us the fossil counterparts of many current animal species, such as the tiger, the dog, the skunk, the viscacha, the tuco-tuco, the cavy, the *Hesperomys*, the deer, the guanaco, the *peludo*, and the *mataco*. My own observations have also added to the many animal species enumerated in the list: man, the capybara, the *Dolichotis* or pampas hare, the *Myopotamus* or nutria, the *Reithrodon*, the *Dicotyles* or Argentine boar, the *mulita*, and the weasel. All that is left is to discover is this land's native genera of bats, and the ferret or *Galictis*, which must necessarily have ancestors as well . . . In time I will discover them, or someone will show me their fossils, and on that day Darwinism will have been handed yet another proof.

I have just likened the animal series to a tree. I am now going to compare a zoological family to a linguistic family. Whether the present-day languages descend from one or from several primitive languages is of little importance here. What is important, what is certain and indisputable, is that languages also transform over time. This is not only proven by the study of languages, but also by history. Nobody would dare deny, without speaking nonsense, that Spanish, French, and Italian are all derived from Latin, or that the latter is not connected with ancient Greek, ancient Saxon, Sanskrit, etc., which in turn points to a common origin for all of the said languages. From the easternmost corners of India, in Asia, to the edges of the Atlantic in western Europe, there spreads a family of languages that are bound together by incontestable resemblances, and derived through successive transformations, largely known, from a common trunk that is now extinct. There was an attempt to reconstruct this lost language. Linguists set to work with unequalled patience, searching for the primitive and common elements present in each of the current Aryan languages, and in this way they have been able to form the ancient language's vocabulary and reconstruct its grammatical forms. The name of that language was not preserved anywhere; it had to be christened, and just like us with our extinct animals, they named that fossil language—a fossil given that it is lost—Primitive Aryan Language. The theory of evolution in the animal series is so certain that a naturalist can go about in the same manner. Making comparisons between the different species of the genus *Felis* or *Canis*, observing the traits that are common to all of them, the primitive organs that today we find more or less atrophied, etc., he can manage to reconstruct a primitive ancestor of the dog, of the cat, and of the other families. And this, gentlemen, is for certain, because the trials that have been carried out have produced results in accordance with the theory. Comparing the different species of the genus *Equus*, made up by the horse, the ass, the zebra, etc., it has been observed that next to the long bone above the hoof they all have two rudimentary little bones called splint bones, one on either side, which seem to indicate the presence of two toes that tried to develop and were unable to do so. They are atrophied organs; and from their presence it was deduced that the primitive ancestor of the horse must have had three toes: and this is true. This type of ancient horse with three toes on each foot lived in the mid Tertiary period, and is now known to science by the name *Hipparion*. The day will come when, in this way, we shall reconstitute and find in the depths of the earth the greater part of the current life forms' primitive ancestors; and then we will be able to reconstruct almost in its entirety the great tree of the animal series, and consequently our own genealogy, along with that of the rest of the current species.

But this is still not all: if transformism is a fact, we can go even farther. We can not only reconstruct the primitive ancestors from which all of the currently existing forms were derived, but also, through simple calculations, predict the discovery of new forms.

Astronomical science today is so advanced and so exact that it can predict the discovery of new stars along with their location. In this way, more than thirty years ago, Le Verrier predicted the discovery of the planet Neptune; this is how the satellite of Sirius was found as well; and in the same way, by induction, Vulcan was recently discovered, between the Sun and the orbit of Mercury.

What I am saying, then, is that in the same way that astronomers predict, though the study of certain perturbations of the Newtonian law of gravity, that between the orbits of planets C and B there must be a new celestial body, so too can a naturalist of the evolutionist school predict, following the Darwinian law of the transformation of species, the discovery of new forms to close the false gaps apparently separating organisms, and can recreate those intermediate organisms that have not yet been found. And this is the clearest proof of transformism's exactness, an exactness that produces more and more evidence placing transformism among the ranks of the exact sciences.

A moment ago I was telling you that the learned Mr. Burmeister is a sworn enemy of transformism, and that he should be considered a precursor to Darwinism all the same because of his *History of Creation.* There was another scholar, no less illustrious, also an enemy of transformism, and who, regardless, was also one of its precursors. His name is Cuvier, the inventor of comparative anatomy . . . the founder of the science that teaches how to classify fossils. "What heresy!" they will say, "placing Cuvier among Darwin's predecessors! . . . " Well such is the case, because Cuvier was the first to demonstrate, and to reduce to laws and formulas, the analogies, the unified plan that animals represent, and because transformism could not have come about without a knowledge of the resemblances that hold between organisms and the anatomical laws that govern them. What was not known to Cuvier, but was nevertheless suspected by his contemporaries Lamarck and Geoffroy-Saint-Hilaire, only to be established as fact by Darwin and his disciples, is that these resemblances indicate higher or lower levels of kinship between the different animal forms. If Cuvier had lived thirty years longer he would have grown weary of trying to use his laws to understand all of the fossils that have come along to fill the gaps in the animal series, and he would have ended up asking himself time and again about the reason behind these resemblances. He would surely have discovered the tie of common kinship between all the life forms of past eras; and the wise man who said: "Give me any bone from the skeleton and I will give you the animal," might have said with equal flare: "Give me at random two different mammals and I will reconstruct the intermediate forms between them." And in fact, if transformism, as all signs indicate, is a reality, the restoration of the intermediate forms is reduced to a rather simple problem: using two known terms to find an unknown one, whose form will be determined by the value of the different anatomical traits in each of the extremes.

For years now I have been intrigued by this system of using numbers to calculate the intermediate forms that we have not yet discovered. I carried out some tentative trials, cautiously at first, and the discoveries that resulted confirmed my deductions. As a result I now have absolute confidence in transformism. Apropos of the edentates, I told you about *Thoracophorus*, an animal whose carapace makes it an intermediate form between the *Mylodon* and the *Glyptodon*. A little bone from this animal's dermis was made known to me six or seven years ago, and applying to it my system of transformist calculation I deduced that it belonged to an intermediate form between the rudimentary carapace of the *Mylodon* and the perfect carapace of the *Glyptodon*. I took my little bone to Paris, where in vain I had one-on-one discussions with several paleontologists about the existence of my supposed intermediate form. They laughed fondly at what they called my transformist delusions. Some time afterwards I was able to study a nearly complete carapace from the animal. I had not been mistaken: it is in fact an intermediate carapace between the groups I had predicted, of which you can see considerable fragments in my collections.

Now guided by these principles, I wrote the following lines in a paper published some seven years ago: "Those islands of Uruguay, Paraguay, Bolivia, part of southern Brazil, and those to the north, west, and northwest of the Republic of Argentina, must have been inhabited by a number of mammals constituting an even stranger fauna than that of the Pampean era. Perhaps those characteristic species of Glyptodonts, Mylodons, Megatheria, and Toxodons of the Pampean era had not yet appeared, *but these species originated from other stranger forms that came before them*, forms that were more different from the present-day species than from the Pampean species, types that lived in the Tertiary period."[2] One of the fossils that most intrigued me at that time was the *Toxodon*, a large mammal the size of a rhinoceros, placed by some among the edentates, by others among the rodents; a few have determined it to be an animal with a trunk more or less like the elephant, and most have placed it among the pachyderms alongside the rhinoceros.

For me, none of these classifications was exact: *Toxodon* cannot be placed in any of the existing orders. It represents an extinct order that shares traits with the rodents and the perissodactyl or odd-toed pachyderms. These two orders are at present separated by a veritable abyss; and in the ancient world there is no known fossil that fills even a part of this void. And so: *Toxodon* is the last descendant of an extinct order that I call the typotherids or the pentadactyls, in accordance with the phonetics of the perissodactyls and the artiodactyls, and because it has five toes on each foot. This order can be placed right between the rodents and the perissodactyls. This intermediate form (I told myself, at all times basing my idea on the rules of transformism about which I have spoken), cannot be isolated, without ancestors or collaterals. It must have been preceded by other stranger forms; and there also must have been other forms to fill part of the gap left between the *Toxodon* and the rodents on the one hand, and the perissodactyls on the other. And I was not mistaken. There were two genera that I only knew by their names, and that fill in part of this gap. One of them is the *Typotherium*, found by Bravard in 1854, which was only recently described, and somewhat incompletely at that. Its most direct analogies are with the *Toxodon*. It fits along with this genus in the order of the pentadactyls, but it is much closer to the rodents than the *Toxodon*. The other is the *Nesodon*, found in Patagonia some forty years ago; I have only been able to study the remains of this genus in Europe. It shares many traits with *Toxodon*, although it is a true perissodactyl, closely related to the rhinoceros, though much closer than this latter one to the former. So there you have two equally extinct forms that tie together the *Toxodon* with the rodents on the one hand, and with the perissodactyls on the other; and the three extinct forms together fill in part of the gap that exists today between this order and that of the rodents. And there are still more forms. I know of, though from incomplete remains, two additional large mammals that are comparable to the *Toxodon*: the *Trigodon* and the *Protypotherium*. I have also seen the radius of a large mammal found near the mountains, which was recently entrusted to me by our President, Dr. Zeballos, in order to classify it, and which is similar to the animal in question but belongs to a different species. I have also seen the fragment of a mandible from another very small animal from southern Patagonia, where it was found by Mr. Moreno, and whose traits also place it in the same order.[3]

Now you see, gentlemen, how many intermediate forms, all of which are extinct, come between the perissodactyls and the rodents, two orders whose current representatives: the rhinoceros and the capybara, are now separated in nature by a veritable abyss. The discovery of such intermediate forms, I repeat once more, can be predicted with confidence, anticipating their traits in advance. This science is far enough along to do so.

Before concluding I am going to cite for you another example taken from foreign soil that was not prophesized, but could have been: the elephant and the mastodon. The first is an existing animal; the second is entirely extinct. The mastodon is a genus very similar to the elephant. The main difference between the two of them is the shape of the masticatory surface of the molars, formed by transversal bands in the elephant and by cone-shaped cusps in the mastodon. The animals are very similar to one another and what is clear is that if the one doesn't descend from the other, then both derive from a common ancestor. It would, then, have been possible to predict beforehand that an intermediate form would be found to act as a transition between the two. This was not done, but the transitional form has been found. A proboscidean has been discovered, which some have called *Elephas Clifti* and others *Mastodon elephantoides*, because no one has been able to determine whether it is a mastodon or an elephant, so intermediate are its traits.

Today the proboscideans constitute a completely isolated order. Their closest relatives are the rodents, of whom they appear to be an enormously over-modified form. But there is an abyss, which surely will be filled one day by intermediate fossil forms, just as the one that separated the very rodents from the perissodactyls was filled, and just as one day we will fill, when we least expect it, the gap that separates man from his primitive ancestor, from which the great apes also diverged. And who knows if these very gaps will also be filled on our own soil! What gives us reason to hope so, in the case of man, is a discovery that I made of the remains of large monkeys in the lower Pampean deposit, and in the case of the proboscideans, a skull fragment from a relatively small animal that a few days ago Mr. Moreno (who for the purpose of these studies hides nothing from me) showed me: a skull that appears to have elephant teeth and comes from a very old formation in northern Patagonia, from where a gentleman whose name I do not recall[4] had just brought it to him. Imagine all the seeming mysteries, in the eyes of the obfuscated anti-transformists, that are bound to dissipate as paleontological research moves forward on our soil!

These discoveries of the fossil remains of present-day species' ancient representatives, this reconstruction of the primitive ancestors of the current zoological groupings, this prediction and discovery of the unknown intermediate forms, all of these facts based on transformist laws, constitute the best proof that can be adduced in favor of transformism and the best wreath that can be offered in honor and memory of its great defender, Darwin. This theory, gentlemen, seems so simple to me, so basic, so logical, so natural, that I cannot understand how there are educated people who fail to grasp it, unless I imagine that the transformist law of intellectual atavism is operating. For me, these transformations and modifications, this existence of numerous intermediate forms, these barely perceptible transitions that lead from one species to another and from the species of one genus to those of another, are all so simple that they seem to me to be within everyone's reach; and so basic, that in relation to transformism I credit myself with no great merits. But when I think back to fifty years ago, at which time Darwin was here having his first ideas about transformism; when I think of how at that time he did not have at his disposition one one-thousandth of the materials that we now possess; and that, in spite of this, after he had conceived his theory he had so much faith in it that he spent thirty years of his life gathering materials before delivering and publishing his first essay, immediately followed by others that were even more voluminous and important; when I remember all of this, I cannot help but admire him; and gentlemen, admire him with me and respect his memory, because Darwin was a great genius and a great scholar.

NOTES

1. TN: *The Cowbell.*
2. OAN: Ameghino, *Ensayo para servir de base a un estudio de la formación pampeana*. Mercedes. 1875.
3. OAN: This fragment, just a few days after my lecture, was given the name *Toxodontophanus australis* by Mr. Moreno; and they christened a fragment corresponding to another very similar animal found at the same site with the name *Interantherium rodens*, without any diagnosis other than that the former is similar to the *Toxodon* and the latter to an intermediate animal between the *Toxodon* and the rodents.
4. OAN: This animal was later given the name *Mesotherium Marshi* by Mr. Moreno, and the only diagnostic he provides to recognize it is that its molars are similar to those of a giant capybara or of a pygmy elephant.

TRIBUTE TO AMEGHINO

(1915)

Leopoldo Lugones

CHAPTER III

His research into the antiquity of man in the Plata region had led him from geology to archaeology, the field to which his first book belonged: *Antigüedades indias de la Banda Oriental* [*Indian Antiquities in Banda Oriental*], which was the product of a scientific expedition to the Republic of Uruguay. His bibliographic knowledge grew as he read papers of a similar nature, but he was aware of its inadequacy as long as he could not contrast his readings with the museums of Europe. To this end, he took advantage of the international convention of anthropological sciences that was to be held in Paris during the 1878 World's Fair, and left for France that same year, with no resources other than a collection of fossils he hoped to sell in order to support himself and publish his already finished book.

His skill immediately drew the attention of the specialists. His talk before the Paris convention and the Americanist convention held in Brussels the following year, his publications in the *Revue d'Anthropologie* and other scientific publications, earned him the title of "scholar," something his own country would not do for some time to come. Such is demonstrated by his book, *Los Mamíferos Fósiles de la América Meridional* [*The Fossilized Mammals of South America*], published in collaboration with Gervais in 1880. What higher certificate than the company of the learned director of the *Journal de Zoologie*? In terms of his own contribution, suffice it to say that of the roughly three hundred species described in that book, there were seventy new ones, all discovered by him. In the same period (1880–81) we have his five studies on the Quaternary bed in Chelles, which have becomes classics of French geology and paleontology.[1]

Thus he repeated the story of Cuvier, who was recognized as a scholar just a few short months after his arrival in Paris in 1785; which is to say, when he was twenty-six years old, just like Ameghino.

The sale of part of his collection earned him 120,000 francs—which goes to show how important it was—with which Ameghino paid for his three-year trip through various European nations, and in the end managed to print his book on the antiquity of man in the Plata region: two volumes of six hundred pages each, with numerous figures. In the meantime, he had gotten married to a French woman, Leontina Poirier, who was to be his loving companion throughout his life. The money from his collection diminished on account of this and other reasons, to such an extent that, upon returning to his homeland, Ameghino found himself in the direst of straits.

To top off his bad luck, the town hall in his home town had barred him from working as an education assistant, because he had exceeded his leave, even though the assistantship was unpaid. In the face of official severity, always inexorable with the useful ones, none of the uncomfortable assistant's qualities were to be of any use: neither the highest mental performance Argentina has ever known, nor the well-rounded studies of a scholar, nor the immense scientific capital that he had brought home with him. Subpreceptor Ameghino was, thus, fired for impropriety (1881), almost at the same time (1882) as Sarmiento made his great achievement as president of the education council. Seven years later, after he had become the brilliant author of *Filogenia* and twenty more papers that were keeping scholars busy the world round, he was dismissed as secretary of the La Plata museum, once again for being a bad employee. Authority always has an infallible knack in these matters; and this is why, since that childhood conflict with the priest in Luján, Ameghino always got along poorly with those who represented the dogma of obedience.

After being left all on his own he started a small bookshop, which he named "Glyptodont Books," not without a certain sarcastic allusion, and decorated the display window with an image of the eponymous monster.[2] He led an incredibly modest life; but he had in his possession the three divine gifts: youth, intelligence and love. Out of this state of abundance would come forth his brilliant piece of work.

Filogenia came into being like so many such creations, in a little room adjoining to the shop, from which Ameghino could hear his buyers' footsteps. The latter were, for the most part, noisy students who got rather in the way and spent rather little. However, in the scholar's heart there prevailed, with the pride of his first triumphs, that optimism of youth; in his brain, reason both fearless and calm at great heights, like an eagle; in the interior of that poor house, the grace of the young French woman who understood his important task and encouraged him with clear-sighted fearlessness; and in this way the work was finished in one year, the work of the happy pauper who, like a bee, needs only a few little flowers . . .

This work set out to provide new laws for the organization and origin of species, rectifying in part the conclusions of Cuvier, Darwin, and Haeckel, whose studies it served, in any case, to complete; just as the first of these scholars had done with Perrault and Vicq-d'Azyr, the true inventors of comparative anatomy. The first application of this science to paleontology, along with the formulation of the laws of phylogenetics, were tasks undertaken, respectively, by Cuvier and by Ameghino between the ages of twenty-six and twenty-eight.[3]

This precocity reveals in both scholars the inventiveness of genius, whose hallmark consists of getting ahead of the results of research and analysis, thus given over to the service of said faculty; and so it is that the genius applies, to things that are still confused and incomplete, the notion of order that he himself has inside of him, which constitutes a revelation of universal truth. Genius is, then, that prescience which understands prior to demonstration and formulates prior to analysis, thus manifesting a clear identity between the laws of human logic and the directions of the phenomenalistic universe.[4] As such the genius amounts to a *deus* or conscious causal force: a creator; and it would seem that this entails for the race in which he appears, either the transcendental predilection of a superior being, or the natural predisposition to organize itself around excellent forms whose prototype is the genius. The corroboration of such a phenomenon carries with it a glorious hope. The appearance of a genius testifies to a race's superior state of vitality and its lofty destiny. In effect: the predetermination that brings it about is a message whose intelligent character seems to me to be undeniable, since the genius's objective is to reveal the truth.

And so that shoot sprung from Italy, which is lush and tenacious, though also sober and sweet, was already rooted in our soil, like the vine that takes the blood of the land into its veins. As there was neither mixing in his lineage nor vice, he took on the qualities of the stock that reaches all the way back to Rome in appearance, so that he himself a Roman did seem. Indeed there is no plant more civilized, nor is any people better suited by civilization. These qualities are nobility, perseverance, generosity, cosmopolitan intrepidity, virile flexibility, the vigorous fortitude of muscle, the salubrious tartness of verjuice, poetry by the shade of a young vine.

(Well that sweet little *piccirilio* (little boy) from Sicily must be gabbling away all by himself there, the one celebrated by Leopardi,[5] who resembles him, and whose life is one never-ending song). Such spirits represent the cordial links of Latinness between both nations; and the man in question brought it all together like no other, with his Italian blood, French wife, and Spanish language. There has never been such a Latin mind: which is to say, one whose bold wisdom was any clearer.

Filogenia begins, logically, with a critique of the systems of classification whose mistake consisted in presupposing a hierarchical distribution of living things, in accordance with arbitrary concepts of superiority and inferiority. The convergence of antagonistic traits in one creature is, on the contrary, rather frequent; and he cites to this effect the case of the *Macrauchenia* of the pampas, who had the neck of a guanaco, the eye socket of a ruminant, parts of the skull and limbs of a horse, the feet of a *Palaeotherium*, the lower jaw of an *Anoplotherium*, and the molar structure of a rhinoceros. The law of Correlation of Growth[6] is not, then, absolute.[7] We are only familiar with scattered and disparate branches of the primitive trunk. However, they do permit us to conclude that the reappearance of lost species is impossible, and that no species can transform into another, no matter how similar they may appear. All of the zoological branches come, in this manner, from one primitive organic trunk, and never one from another, as Darwinist transformism believed. On the contrary, they evolve in progressive divergence from one another. This double chronological principle was in and of itself an original idea. It entailed another idea which was even more important, within the transformism to which it ascribed, and that as a fundamental law of evolution is, indeed, unassailable—the conclusion that man is isolated on the species map, and that no monkey, even a hominid, can transform into a man. He believed "that the different groups of animals are perfect in and of themselves, *as long as their disposition allows them to keep their advantage in the fight for life*,"[8] which means that there is no qualitative hierarchy between them. But there is an unfavorable sign that can lead to the disappearance of the species: the ossification of the skull, which, impeding the development of the brain, constitutes a veritable "process of bestialization," a certain prelude to death. In sum: triumph in the fight for life depends upon the brain's development, an idea whose spiritualist beauty sets *Filogenia* apart for its high philosophical relevance as compared to the other works of its ilk. We are equally far from the miserable Biblical prejudices that disparaged Cuvier and Darwin's scientific method, and the systematic bestiality of Huxley and of Haeckel. The consequences of this philosophical orientation would quickly become clear in the scholar, whose *Credo* finally faced, resolutely, the problem of immortality.

The development of the brain, which diminishes the size of the maxilla, also reduces the number of teeth which tend to fuse together; and in this way beastly gluttony gradually becomes less intense. This second consequence also ensures success in the struggle for life due to a reduction in mass. When Ameghino set forth these principles forty years ago, physics had still not discovered that under certain conditions a decrease in matter is capable of generating the most powerful and subtle strength.

From these ideas emerged his "dental law," which, while distinct from the classic law perfected by Cuvier, reveals how Ameghino, too, assigned a fundamental importance to the masticatory organs.[9] His two masterpieces are, in effect, *Filogenia* and the paper "*Sur l'evolution des dents des mamifères*" (in French) ["On the evolution of the teeth in mammals"]. In favor of the aforementioned law, which he developed in a masterful chapter on dentition (Chapter X of *Filogenia*), he made surprising discoveries, such as the *Peltephilus*, which he foresaw fifteen years before anyone else. It was a creation that, even on its own, would be enough to secure a scholar's fame.

Ontogenesis, or embryological recapitulation, also seemed to him to be insufficient to determine living things' genealogy, since many ancestral traits disappear due either to elimination or synthesis, and only show up again sporadically if at all, while others only reappear in the full-grown animal.[10] And so it happens, in both cases, with the teeth, which, according to the principle of recapitulation, should be stronger and more numerous than those of an adult in the human embryo, as the ancestors of man had a more abundant and powerful dentition; while at the same time the wisdom teeth do not emerge until after adolescence, and progressively later still at that.

The procedure that *Filogenia* establishes for genealogically determining and classifying species is seriation, or the arrangement in chronological order of the groups whose common traits one wishes to study, for which one need only apply the phylogenetic laws that the book sets forth in a system of precise subordination. This law is composed of fifty-some principles and corollaries, which, it is safe to say, can be reduced to roughly a dozen summarized laws, whose overall goal is "to find, by means of two known terms, one unknown term whose form shall be determined by the values of the different traits at each of the two extremes."[11]

Indeed, the entire system of *Filogenia* tended to express zoological laws in terms of mathematics: another point in common with Cuvier, who saw the Correlation of Growth by which some forms determine others *"comme l'equation d'une courbe entraine toutes ses proprietés."* Our wise scholar, who was a bit less categorical,[12] complemented his work on dental formulae with work on digital formulae, and thus found the hand to be an organic prototype nearly as important as the head. The primitive formula had been the same as in man (5/5), which once again undermines the supposed general applicability of the

Correlation of Growth, as it so happens that animals as cerebrally inferior as the *Typotherium* had hands very similar to those of man.

The far-reaching consequences he saw in the constitution of this "mathematical zoology" is evident in the fascinating comparison he made between said science and the field of cosmography, and also reveals, more than anything, the depth of his philosophical spirit. Within the harmony of the universe, the laws around which the cosmos is organized must be, in effect, analogous to those of our own biology;[13] and it is a well-known fact that mathematics constitutes the underlying instrument of astronomy. To subordinate biology to mathematics as well is to accept outright that the precision of human logic can satisfactorily formulate the harmony that holds between all phenomena, which is to say the state that we call the universe, thus defining this immense conglomeration through the notion of the human spirit's precision. Man is the microcosm of the macrocosm, the human mind is the mirror of the universe, as ancient wisdom has already said, and it is curious to see that modern science has come to the same conclusions. Understanding unity as harmony is the key to the universal secret which falls upon the genius to unlock; and man's transcendental ontogenesis is the revealing phenomenon that thinkers throughout time have described using mathematics. And mathematics—to use opportune Pythagorean terms—by demonstrating the unity of phenomena, show the human spirit's noumenal character: the ancient *logos*, thus called because it formulates the laws of the universe. The unity of all matter, which was one of alchemy's postulates, has been transformed into a scientific finding. The day is not far off when the existence of the gods will be rediscovered, or rather the noumenal transcendence of the forces of nature, and then we will understand the accuracy of the Ciceronian maxim: "It must be agreed that nature is endowed with reason."[14]

Lastly—though the book is so interesting that we will have to continue examining it, as we are far from having worn it out—the supposed gigantism of primitive species is likewise false. On the contrary, species evolve in the opposite direction, and when they grow so large as to upset the equilibrium of their surroundings, which is particularly characterized by difficulty in finding sufficient sources of food, they disappear. In this way he was able to predict twenty years in advance the discovery that proved that the glyptodonts had descended from the armadillos. As I see it, the idea of primordial gigantism originates from the fact that since the remains of large animals are better preserved, they were also the first to be discovered; to which we must add the simultaneous existence of a Biblical prejudice.

This was foremost demonstrated by the discovery of the homunculus, a Patagonian member of the Lemuridae family whose bodily structure anticipated the appearance of humans with astonishing perfection. This find confirmed *Filogenia*'s most revolutionary proposition: man was not the descendant of pongid monkeys, nor were the pongids the descendants of mankind, which would in turn have no known ancestor. Rather, the ancestor of said monkeys must have been an animal more similar to man than they themselves, in terms of certain anatomical details.[15] As such, "it is not man who constitutes a perfected monkey, but rather, on the contrary, it is the monkeys who seem to be men turned into beasts."

Let us return, then, to the original purpose of his writings, which is another sign of philosophical nobility: the origin of man, a question on which the emancipation of many a spirit rides.

His research into the antiquity of man in the Plata region was to produce the most remarkable of discoveries, the boldest of findings, the most startling developments in his scholarly life. This is how, apart from paleontology, he explored with equally brilliant might the fields of geology, paleogeography, anthropology, and archaeology.

The discovery of human fossils in the horizons of the Quaternary—a demonstration that earned him attacks from both the scientific establishment (Burmeister) and clerical circles—was later followed up by the corroboration of man's existence in the Pliocene and in the Miocene, which of course brought with them the same results.

It has been said that Ameghino was systematic, in the sense that he made his discoveries fit into the mold of his theories. We will have a look at what truth there is in this. For now let us just say that the confirmation of humans' existence in the Miocene contradicted one of his hypotheses regarding the origin of man if regarded as a descendent of the homunculus. This is because the homunculus of the Tertiary, which is the one most similar to man, would in that case be his contemporary; and although the homunculus appears to come from the Eocene, the works of man in the Miocene reveal an intellectual (use of geometric solids) and organic (the perfection of the hand) development that must have arisen in a prior epoch.[16]

And here I must raise a fundamental objection, which to my mind originates in the very same phylogeny set out by our dear scholar. He established, thirty-three years ago, that man existed at a time when the present mammals had not yet appeared; that he did not descend from a monkey; that his brain is not superior to that of the animals quantitatively but rather qualitatively; and that his structure has hardly varied since the distant past. His most basic law, which explains species' progress and decline, subordinates biological fitness to the development of the brain.

However: the stability of an organism as functionally complex as our own implies that by the Pliocene man must already have had a very long history; and man's crafts in the Miocene demonstrate a cerebral development that is indicative of advantages in biological fitness: the same ones that hold at present. In addition, in order for the brain's quality to have become matched up with and then have overtaken its quantity, an equally long process of selection for intelligence would have to have taken place. This would indicate that in the Pliocene man already had quite ancient intelligent predecessors. Man's digital formula also turns out to be of the most primitive or primordial sort, while also being of the most advantageous of all structural formulae for those who possessed it. Lastly, the brain's tendency to grow is shared by all animals, and appears to be a hereditary advantage.

Since it is clear that different organs have developed in different directions and at different paces in accordance with the way in which organisms have used each of them, the preservation of an organ since primitive times, such as the five-fingered hand, indicates the antiquity of the organism who possessed it. At the same time, if, in the same individual, there exists another highly evolved organ, such as the brain in the case of man, this would demonstrate the individual's dominant and progressive tendency toward using the organ in question. Related organisms do not evolve, according to Ameghino, towards the subdivision of their bodily structures, but rather to the contrary. Their pieces, which are quite numerous at the outset, slowly come together into more synthetic ones, which simplifies their structure, but increases its functional complexity. In the case of an organism as structurally complex and old as our own, the remoteness of its origin is clear. This, in conjunction with our organism's remarkably advanced cerebral evolution, allows us to surmise that intelligent creatures analogous to us, if not of the very same human type, could have inhabited the earth since the beginnings of the creation of life. If, furthermore, a tendency toward cerebral growth is hereditary in all animals, this points to a common ancestor for whom this developmental tendency was dominant; and since no organ develops without the exercise of the function that belongs to it, this tendency in the hypothetical organism in question must have manifested itself in activities involving the intellect.

If we turn to the ontogenesis of the vertebrates, we will notice that the proportion of cerebral volume is much larger in the embryo; that the first fold of the embryonic tissue,[17] where the layers of the blastoderm will develop, is the cephalic fold; that out of the first of these layers to appear are born the brain and the medulla;[18] and that the first instance of ossification[19] takes place in the skull. Any ablation of the cerebral region has significant consequences for the organism, the most common of these being the interruption or retardation of growth. In such cases, the growth of the four limbs can also be suppressed,[20] along with organs as important as an entire kidney or lung, while the brain itself is unaffected. Originally the spinal column was, as can be seen in the fossil proto-vertebrates, a cartilaginous dorsal cord, a sort of sheath that encased the medulla, which already existed. All of this provides us with the grounds to surmise that the brain and the cranium were not prolongations of the spinal column, but the other way around; just as the first living thing must have consisted of a sort of simple brain-like ganglion.

According to Ameghino's biological doctrine, this organ's potential would constitute the fundamental condition for the progress of the species, and the consequence of this organ would be the evolutionary plan which this very progress amounts to. From this it follows that man, as the highest cerebral entity, would turn out to be the first living thing. Originally, he would have been nothing more than a brain, which would have gone about creating the material organisms or "bodies" that he could use as tools and support. When I say "man" I do not refer to the current human organism, but rather to the intelligent organism, which, as such, would in effect be our most direct ancestor. Modern biology already believes that the formation of the skeleton is subordinate to that of the soft organs; and an embryo's cranium passes successively through membraneous, cartilaginous, and bony states, which must be ontogenetic. In fact, the embryonic brain's first capsule is reminiscent of the basal ganglion in fish, a membrane which, according to Jakob, "is substituted in the higher vertebrates by the cerebral cortex."[21] As such, human intelligence would have presided over the whole of life's evolution on earth, probably in an analogous form to that formerly and presently used by this much larger organism's intelligence. Yet at this point it would be useful to clarify the biological and philosophical value of these terms a bit.

Science has demonstrated that all living things are more or less intelligent. This is to say that they understand and try to understand the scheme of life in relation to their individual and specific prosperity. Insects and birds provide examples of this, all the more incontestable given their independence from our own biological design. Thus, barring ants and bees, which possess the rudiments of a central nervous apparatus, the rest lack such an advantage, which does not keep them from giving us remarkable proofs of an intelligence analogous to our own; for were it otherwise we would not appreciate them. If the earth is also a living thing, and even the foremost of such things as far as we are concerned, it must possess an intelligence on par with its complex functions, that is to say, the same faculty of understanding that we and the insects have. And we should keep in mind that the life of humanity makes up part of the life of the earth. What we call "creation" would in such a case be this very life at work; and if, as we see in all living things, this requires the vital faculty that we call intelligence, the evolution of said life could not have been purely mechanical.

Let us recall that for species, the problem of duration in time consisted of a double and divergent command over the vegetative and mental functions, by which it turned out that the former gave rise to a progressive material growth, or a "process of bestialization." This in turn brought about an excess of body mass and consumption of foodstuffs, leading to an imbalance with the environment, the ultimate cause of extinction. This bestial progression is characterized, we have not forgotten, by the development of the mandibles and the resulting reduction of the cranium, whereas in the opposite scenario it is absolutely to the contrary. The expansion of that capsule, which retained the primordial ganglion's spherical form, in all probability is what made the bipedal state necessary. It thus gave it the sturdiest means of support, due to the natural gravitation of the cranial weight atop a pillar: the vertebral "column" strictly speaking. This actually becomes an axis in other types of creatures, especially the quadrupeds whose heads are essentially suspended from the neck, which in turn needs to be proportionally more robust. By virtue of Geoffroy Saint-Hilaire's law of compensation, the maintenance of this robustness takes away from the development of the cranium, creating a vicious circle of bestialization. The most advantageous position for the brain to grow without impinging upon balance is, then, that of the human skull; and this happens to be particularly visible when man abandons his normal state and environment and enters the water. He does not float naturally like the rest of the animals, but rather sinks, unless he creates a special equilibrium with the compensatory movements of swimming, since his head is heavier than the rest of his train.[22] If the human organism was primitively aquatic and amphibious, as many modern biologists believe, the brain's development could have been one of the causes for its more modern preference for dry land. In any case, this environment, in turn, evidently demands a life of more complex relationships; and with this, an increase in cerebral activity (Jakob. ob. cit. I, p. 13). Moreover, the brain's evolution, in terms of the growth and diversification of its mass, took place in the form of a backwards rotation, which, given the natural resistance of the condylar insertion and of the neck muscles, would have tended to displace the axis progressively towards the *norma verticalis*. The corresponding reduction in prognathism, corroborated by Ameghino, as I already said, strengthens this hypothesis.

The cranium, then, would have been, for animal evolution, the primitive organ or organism, a case, which was first made up of a large number of pieces, where the cerebrum was housed. The Wormian or supernumerary bones might say something about this; especially if they were compared with the skull pieces of the fossil cyclostome and ganoid fish, and with the present-day plagiostomi, in order to examine the other material extreme. The vertebral column would thus have stemmed from the skull, summarily repeated for each vertebra.[23] Likewise, the study of these pieces and of the hemal arches allows us to conjecture that the first living thing was cylindrical and fusiform like a worm. A cylinder-cone developed out of the great cerebral cell, which was spherical like the earth,[24] and some of whose papillae would have elongated into sensory appendices that would have turned into limbs. In this way the same plan would apply to all living things through the following geometrical relationship: sphere, cylinder-cone, rectangular parallelogram, and parallelogram of forces, represented in a series by the cerebrum, the medulla, the thorax whose bases are the pelvis and the shoulders, and the peripheral action of the four locomotor limbs. This mathematical biology brings together organic and mineralogical creation, which are daily proven to have more and more similarities. The proto-animal would have vegetated, following a course analogous to the growth of plants, whose prototypical ontogeny would in turn be found in the osmotic products of colloidal precipitates.[25] The shortcoming of biological theories is that they have only considered the scheme of life from the perspective of the animals, whereas in nature everything is interrelated, and so this scheme should be found foremost in mineral forms. In this way the scheme of zoology would turn out to be a part of the biology of the earth, which is, to our mind, the primordial being.

As for the rest, mineral life has been in existence much longer than organic life. The latter was preceded by the former, and if it did not arise from it, it is at least certain that it lives off of it, and returns to it once it decomposes. All of this allows us to safely surmise that mineral life is the prototype that determines organic life. The former's mathematical organization is readily apparent because it is less complex and more stable; in light of which Ameghino's mathematical zoology takes on a new and more peremptory confirmation. His law of cerebral development, a necessary condition for species' progress, gains ground as well. The scheme of life would in fact be determined by this need. Each organ thus carries out its function for a time and then disappears, making way for another more advantageous one; and so the whole of the organic process consists of a progressive acclimatization of the organism to the work of the brain, which is the beginning, middle, and end of animate existence. Everything, from the monad to man, must progressively grow in intelligence in order to subsist; and under close examination, the task of adaptation itself is no different. The life form that best understands its prospects in any given environment also takes better advantage of them, and subsists.

I will add that when I said that animal creation was "presided over" by human intelligence, I only meant to refer to the preeminence of the intelligent proto-organism whose most direct descendants would be us, mankind. The human organism, in the different forms that it must have assumed in all the different geological environments (molluscoid, cartilaginous, vertebrate) would in each case be the oldest member of each order. Because of this and its higher intelligence, it would have held an influence over the development of the other organisms. These organisms would have been, then, later and less fortunate adaptations to the work of the brain; and according to such a hypothesis, man would have been, temporally, the first of the vertebrates.[26]

NOTES

1. OAN: G. et de Mortillet, *Le Prehistorique*; passim.

2. OAN: The shop was located at what is today 2339, Calle Rivadavia.

3. OAN: Linnaeus published his renowned system of classification when he was twenty-five.

4. OAN: In his "essay" about Darwin (*Filogenia*, LVII), Ameghino spoke highly of the English naturalist, a man of genius and a great scholar, saying: " . . . Fifty years ago (this was in 1882), at which time Darwin was here having his first ideas about transformism . . . he did not have at his disposition one one-thousandth of the materials that we now possess; and that, in spite of this, after he had conceived his theory he had so much faith in it that he spent thirty years of his life gathering materials before delivering and publishing his first essay, etc."

5. OAN: *Il Passero Solitario*. I Canti, XI.

6. EN: The principle that certain traits only occur in conjunction with other seemingly unrelated traits, such as possessing mammary glands and having two condyles on the occipital bone.

7. OAN: So as not to stray from American topics, I will recall that Cuvier, in his cited essay on the *Megatherium* (Ann. du Mus. d'Hist. Nat. T. V.), had assigned to said animal a relatively small tail, based on the small size of the posterior side of the sacrum, pointing to the smaller size of the organ in question. However, the *Megatherium* had, on the contrary, a remarkably developed tail.

8. OAN: Italics mine. The forgoing statements of Ameghino show, on the one hand, his proximity to Geoffroy Saint-Hilaire, who first proclaimed in his Anatomical Philosophy, the original unity of all species; while, on the other, the idea of their proven autonomy shows an affinity with Cuvier as well. This is the best proof of his honor and his independence.

9. OAN: Not just because they are the most persistent fossil remains, but moreover because of their constant activity towards preservation of and prosperity in life, which is the fundamental cause behind their refinement, as animal breeders know all too well—"breeds improve through the mouth"—they entail an intense evolutionary process.

10. OAN: In his cited work on dental evolution in mammals he defines it much better (p. 36): "Embryological development is indeed a recapitulation of phylogenetic development; yet, in order for the traits to be present and visible in the embryo they must have been functional during the life of the individual or have appeared in extrauterine life."

11. OAN: I have slightly corrected this clause, which would otherwise be slightly faulty in terms of the complementary clause required by my text.

12. OAN: Cuvier himself duly added, when describing the consequences of his law, "Mais quand il s'agit de l'appliquer, il est un grand nombre de cas ou notre connaissance théorique des rapports des formes ne suffirait point, si elle n'était appuyée sur l'observation."

13. OAN: Van Tiegen once said that if the earth is a tiny part of the universe, then its vegetation can also be considered a tiny part of the universe's vegetation.

14. OAN: Marcus Aurelius expressed this philosophical conception of the world sublimely in his *Meditations* (XI, book IV). "View it," he says, "as an animate being, composed of just one material and just one soul. Observe how everything conforms to just one feeling; how everything is carried out through its influence alone; and, lastly, just how evident the solidarity of all things is." Platonism had the same idea of the world, understood to be the proto-animal, and the teaching of the Eleusian mysteries.

15. OAN: In particular the vertebra of the coccyx, the incisor bone, and the astragalar foramen.

16. OAN: In the Upper Pliocene man's skull was essentially the same as in the present day.

17. OAN: The formation of the embryo consists of a series of folds and invaginations that eventually make up the fundamental framework of embryonic life.

18. OAN: In a chicken embryo, for example, the eye begins to form within the first hours of development.

19. OAN: Between the fortieth and forty-fifth day of a human embryo's life, the areas exhibiting ossification are the parietal bones and the occipital and the superciliary protuberances. Between the twentieth and the thirtieth day, when the embryo is hardly five to ten millimeters long, the head is already taking shape, and the eyes, nose and mouth are detectable.

20. OAN: The pituitary gland, which is one of the physiological apparatuses employed to maintain circulatory pressure, and with it the harmony of vital dynamism, has a direct—perhaps preponderant—role in organic growth, as its disruption gives rise, it would seem, to cases of gigantism. Let us add, under the pretext of an important phylogenetic proof, that in the embryo and the adult this gland represents the ancestral mouth (paleostoma) of the vertebrates (*Robert Legros, Developpement de la cavité búchale de l'Amphiouxux lanceolatus. Anat. Micros., t. I, 1898*). And lastly might we add that in this area of the brain we find the pineal gland, or the rudiment of a third visual organ, which has become a true "pineal eye" in certain proto-vertebrates and reptiles.

21. OAN: *Atlas del Cerebro de los Mamíferos de la República Argentina*, by Ch. Jacob and Clemente Onelli, Chapter I, p. I.

22. OAN: Verticality, limited moreover to the neck, in the bipedal state of birds arises from the fact that this balance is an adaptation, which in turn is the result of the transformation of the forelimbs into wings, after which they were no longer of use for locomotion on the ground. In the kangaroo the bipedal state must have arisen in order to maintain the safest position for its contents, i.e. with the maternal pouch facing up. In the jerboa it must be a persistent marsupial trait; as Ameghino demonstrated in the paper on "*El Diprodontes . . . y el origen de los roedores, etc.*" ["The Diprodonts . . . and the Origin of the Rodents"], the rodents descended from the primitive marsupials so directly "that the plagiaulacids (*) are the rodents of the Mesozoic era." I myself will add, as coinciding proofs, the tendency of all rodents to stand up on their hind legs, the persistent bifurcation of the uterine horns that tends to be common among the females, and the premature state in which some of its smaller species are born. Moreover, it is clear that the bipedal state of the kangaroo and the jerboa is not the product of cerebral or cranial development, which in fact is scarce in these animals. Lastly I will add that these ideas regarding the origins of the bipedal state do not belong to Ameghino. He attributed it another, completely different origin: the need to stand up in the high grass to see off in the distance, a stance that would have become habitual through repetition. He saw in this, to the contrary of what I believe—even though my proposal agrees with his phylogenetic laws—the reason behind the advantageous development of the human cranium. The *Homo caput inclinatus*, one of his most remarkable discoveries, appears to confirm said belief. However, it is fair, all the same, to note that the bipedal state of the kangaroo has not influenced the cerebral-cranial growth of these mammals, even though they figure among the oldest ones on earth. I think, moreover, that the need to raise oneself up would not have brought about the bipedal state, but rather the lengthening of the neck, as in the giraffe. All of which does not change the fact that, as to all the rest, I owe the entire scientific foundation of this digression to Ameghino.

(•) Because the diprotodonts belong to the order Plagiaulacidae.

23. OAN: In various classes of vertebrates, some of the embryo's vertebral segments fuse with the primitive skull, thus coming to form part of the bony capsule (J. B. Johnston, *Nervous System of Vertebrates*, Chap. III, p. 65. See also p. 72, loc. cit., where he summarizes the vertebrates' "process of cephalization").

For those who wish to study in more depth this fundamental matter, regarding the hypophysis or pituitary gland, which is one of the most interesting points, see the initial and classic work by van Wijhe, *Uber die Mesodersegmente and die Entwickelung der Nerven des Selanchierkopfes*, Amsterdam, 1882. He was the first to study in the so-called selachians, the relationship between the "somites" or embryonic proto-vertebrae of the neck, and the nerves and muscles of the gills and the tongue. See also Lubosch, *Vergleich anatom, Untersuch uber den Ursprunf und die Phylogen, des Nerv. Access. Willis*, Arch. f. mik. Anat. Bd. 54, a. 1899. Willis's accessory nerve is, as you will recall, the elongation of the bulbar root of the eleventh pair of cranial nerves. For the general phylogenesis of the nervous system in the vertebrates, one can consult the approximately one hundred and fifty or so works that, as far as I know, comprise the list of works on the subject: Balfour (F. N.) *Comparative Embryology*, 2 vol, 1885. P. Ramón *Investig. microgr. en el encéfalo de los batr. y rept., 1894*, Huber (G. C.) . . . *Minute Anatomy of the Sympathetic Ganglia of the Vertebrates*, in Journ. Morph., vol. 16, a. 1899. Elliot Smith *On the Morph. of the Brain in the Mamm.* Trans. Linn. Soc. London, ser. 2, Zoof. 8, a. 1903. Ramón y Cajal, *Textura del sist. nerv. del hombre y de los vertebrados*, a. 1904. Jakob y Onelli, op. cit., which is the latest in its genre and is particularly recommended as far as our interests are concerned, as it deals only with the mammals of Argentina.

24. OAN: The brain's cortical cells are ordered, according to Jakob (op. cit. p. 28), "in the shape of superposed horizontal layers (principle of stratification, analogous to the earth's geological layers), etc." This erudite histologist is not, by the way, at all fanciful; and since he is, moreover, an irreproachable materialist, there is no reason why the same shared *peccadillo* cannot be forgiven of a pitiful spiritualist like myself. I will add that the fact that a bipolar electrical element forms the cell (Gallardo, An. Mus. Nac. Ser. 3rd, v. VI) provides another analogy with our planet, while also explaining, perhaps, organisms' double structure.

25. OAN: The cell is a mixture of colloids. Ameghino, in his unpublished notes printed by the magazine *Physis* (#3, p. 15), considers living things to be "machines in movement, built with colloidal materials." Gallardo believes that cellular division, which is the very origin of organic growth, is by nature an electric-colloidal phenomenon (loc. cit. p. 259).

26. OAN: My truncation of this hypothesis is only due to the need to cut off a digression which has perhaps been excessive. However, I am neither shocked nor daunted by the idea of intelligent life forms analogous to us, and whose organism was molluscoid or pisciform. The insects of the present day act like us in many cases, and at times with more intelligence; which proves, moreover, that this faculty does not depend on different kinds of organic development, but rather that this development, in all its many forms, aims to serve said faculty. As for the rest, the Nagas or divine mollusks, and Dragons or fish-men of the theogonies are there to justify me. Who is to say that they are not memories, like the dragons, now recognized by paleontology? To imagine is, in short, to relive prototypical impressions and memories. And does the entire *modus operandi* of natural selection not rest upon hereditary memory? [. . .]

NATURALISM AND EDUCATION

Lecture given before the Catholic Club on August 21st, 1880

José Manuel Estrada

[. . .]

Gentlemen:

The schools are just fine today. The concept of the supernatural is to be either accepted or rejected. If it is accepted, one must acknowledge man's intellectual decadence and his rehabilitation through Christ. If it is rejected; either it is affirmed that man is, in his present state, an intrinsically imperfect creation of God, which is absurd; or it is agreed that he is a fortuitous or inevitable product of material and force, born of virtually life-giving physico-chemicals, and by the same token is destined to dissipate into nothingness. I will leave aside abject materialism and face the naturalist school, disseminated by apostasy. Those of you here cannot waver, since you have the honor of calling yourselves Christians. Naturalism denies about man those things of his that affirm revelation. Ergo it does not know man; and since it does not know him, it is unaware of that concept which is the starting point of all philosophical speculation, and the absolute basis of all transcendental relationships, the study of which makes up the substance of philosophy. Ergo, naturalism is not a philosophy. Do not waver in acknowledging this with me; and respond to this question without delay: is it possible to devise a fruitful education system that is based on naturalism, whatever may be the doctrinal accidents that it reviews: the arrogant narrow-mindedness of Comte, the insidious moderation of Stuart Mill, or the elegant indolence of Victor Cousin? . . . As for me I do not believe so, and I am going to unceremoniously give you my reasons.

Man is by nature ignorant, and through his degradation inclined toward evil.

Naturalism concedes the first proposition and denies the second; but the first still needs to be explained, albeit briefly, as we are all in agreement in this regard.

St. Thomas teaches that the human generations would have been ignorant even if they had been innocent; but he adds that their innocence would have leveled out the path, now craggy of science. It seems to me that the great master's explanation is as satisfactory as it is simple.

The powers of the mind are exercised and increase by means of organic conditions, purely instrumental to be sure, but necessary for the system and for the unity of human nature; from which it follows that if, by hereditarily preserving the primitive form of justice, he had preserved the subordination of matter to the spirit that was bound to and followed from him, his intellectual progress would have been quicker and vaster as compared with today in his degenerate state.

With these qualifications and keeping in mind the observable state of humanity, the verdict is beyond serious debate. The exceptions confirm the rule. The prophet who sees in the future what God wants to announce to men: Moses, studying cosmogony and geological evolutions with sublime plainness in order to safely instruct the humble and astonish the sincerely wise; St. Thomas, giving an immortal form to the science of God and of man, a science imbibed in contemplation before the cross; they only reveal, gentlemen, the intellectual magnitude that man could have possessed if he had preserved grace, and that he could humbly expect if he had a faith nary the size of a mustard seed. Yes, gentlemen, man is ignorant; and to diminish his ignorance in order to broaden his domain, to acquire beam by beam the radiance of truth in order to mitigate his servitude, are the pressing duty and rigorous task that God imposed upon him, kindling in him the most noble desires, and proclaiming them from the sacred lips of the Redeemer. We should love science. Christ has not come to this world to abrogate the law that compels us to search it out. Naturalists and Christians coincide.

Nevertheless, naturalism cannot give science to the world . . . I can foresee a response and I will gladly concede all that serves as its foundation. Making do without revelation, and even rebelling against it, the man of the nineteenth century has dug deep into the heart of nature, has investigated laws heretofore unknown, and has given them marvelous applications. Physics and chemistry are sciences of this century. I recognize this and pride myself on it alongside my contemporaries and reject, as unproductive, the argument that I could espouse in order to retake the place of honor corresponding to the multitude of Christian scholars in this progress. Natural history's expansion is enormous . . . I won't be the one to deny it, nor even cast doubt upon the rare observational skills of men so rash when persuading like Darwin; nor will I be the one to fend off their villainous consequences and disdain the indecent trickery of Haeckel and the credulous and ridiculous pride of those who discovered in a grain of phosphate of lime the germ of the primordial anchovy. This century has created philology! It is also true that this is its own merit, and as far as I am concerned, I won't insist upon pointing out that the richest collection of information at its disposal is the work of Christian missionaries, that the most distinguished of all the thinkers that have shaped their method, Bopp, was a faithful son of the Holy Church; nor even (which we might see as an imprudent renunciation of logical advantages), that the most solid results of philology's inquiries,—like the proof of languages' interrelatedness, and they shorten more and more the number of branches into which these were believed to be divided, and group them into families ever closer to the unity of the trunk the more extensive that they grow—provide imprudent but conclusive confirmation of Mosaic teaching, which was the object of laughter for the ignorant henchmen of that fatuous comedian by the name of Voltaire. I concede it all, gentlemen; the merits that the century deserves, and those that it usurps; those of its men, and those of ours; and I concede to it even the merit of confirming truth itself, because in the end I am a Christian, and to he who wishes to take from me my tunic, I must leave my robe as well.

Yet here is where I cease to concede, and begin to judge. From all this to which we have consented, people commonly deduce naturalism's ample capacity to illuminate the world through science, and consequently, the falsity of the proposition that I uphold. Nothing could be farther from the truth, gentlemen. The sciences of observation, whether they belong to the material order like chemistry or to the moral order like philology, classify facts, define phenomena, perhaps formulate their immediate and secondary laws; but they are powerless to discover the overarching connection that binds them, within their metaphysical conditions of production, to a universal harmony, subject to one sublime law. Counting, weighing, and measuring are clearly scientific techniques, but they do not exhaust the scientific program; and expand though it may the lot of things that man grasps by weight, by count, and by measurement, and though expanding it may enrich the material for scientific inquiry, the horizons of science will not have been widened, for science seeks primary motives and primary causes, the ultimate purposes and the ultimate destinies of things and of beings. If man's ignorance consisted merely of his powerlessness to appraise phenomena and their conditions, naturalism would be sufficient in order to gradually dispel it. But neither from the mind of the Christian, nor from the mind of the atheist logician, nor from the spirit of anyone who has raised himself an iota above the state in which, through an excess of our primordial degradation, pure animalism and barbarism are almost indissolubly intertwined, will there ever disappear—though all the curiosities of the visible world were dried out, and the circumstantial reasons behind all experimental facts were hidden from it—this other curiosity: Who am I? Nor this other one: From whence do I come? Nor, lastly, this distressing problem at the center of faith's sweetness and of the piercing anguish of disbelief or of doubt: To where am I headed? . . .

It is evident that naturalism does not respond to such tremendous and spontaneous questions: it is evident that it disdains them and relegates them thus to mere fever dreams, scorning as soothsayers all men who are tormented by them, even more harshly in the case of those who are consoled by them, treating with terrible sacrilege Christ who inspires them! . . . No, have no doubts about it: naturalism is incapable of extirpating the ignorance that is native to man; and this is why, even if we consider education from a strictly intellectual point of view, naturalism lacks the resources to further it by setting it on a safe and fertile path.

[. . .]

YZUR

(1906)

Leopoldo Lugones

I bought the monkey at a public auction for a circus that had gone broke.

The first time I toyed with the idea that would lead to the experience related in these lines was one evening, reading I recall not where, that the natives of Java attributed the lack of articulate speech in monkeys to abstention, not inability. "They do not speak," they said, "so that they will not be put to work."

This idea, of little importance at first, would in the end become so absorbing that it turned into the following anthropological postulate:

Monkeys were once men who for one reason or another stopped speaking. This led to the atrophy of their vocal organs and of the brain's language centers; it weakened the relationship between the two almost to the point of suppressing it altogether, transforming language in this species into an inarticulate cry, and so the descendants of the primitive human became animals.

It is absolutely clear that if this could be demonstrated, it would explain all of the anomalies that make the monkey such a unique creature; but there could be only one possible demonstration of this: to give language back to the monkey.

In the meantime I had traveled half the world over with one of my own, drawing him closer and closer to me by means of episodes and adventures. In Europe he drew people's attention, and, if I had wanted, he would have obtained the celebrity of a *Consul*; but my businessman's seriousness could hardly be reconciled with such stunts.

In working away at this set idea of monkey language, I used up all the bibliography concerning the problem, with no results to show for it. I only knew, with total confidence, *that there was no scientific reason why a monkey should not speak*. This came after five years of meditation.

Yzur (a name whose origin I never did discover, since his former owner did not know it either), Yzur was truly a notable animal. His circus training, even if it amounted almost entirely to mimicry, had certainly developed his faculties; and this was what drove me all the more to test out my apparently absurd theory on him.

Furthermore, it is known that the chimpanzee (Yzur was one) has the best brain among the monkeys and is one of the most docile, which increased my chances. Every time I saw him move along on two feet, with his hands behind his back to keep his balance, walking like a drunken sailor, my conviction as to his arrested humanity grew stronger.

Truth knows no reason why monkeys should not enunciate at all. Their natural speech, that is to say, the system of cries with which they communicate with their fellows, is quite varied; their larynx, as different as it may be from ours, is not so different as that of the parrot, who speaks regardless; and as far as its brain is concerned, apart from the fact that a comparison with this latter animal will dispel all doubts, one need only remember that the idiot's brain is also rudimentary, in spite of which there are cretins who manage to utter a few words. In regards to Broca's gyrus, it depends, true, on the complete development of the brain—the fact that it has not been proven that this is *necessarily* the site where language is located aside. If it is anatomy's best case for localization, the facts to the contrary are themselves incontestable.

Happily, monkeys have, among their many bad qualities, the desire to learn, as their penchant for mimicry attests, a felicitous memory, an ability to reflect which gives rise to profound cunning, and a strong attention span, comparatively more developed than that of a child. It is, then, a very favorable pedagogical subject.

Mine was young moreover, and it is a known fact that a monkey's youth constitutes its most intellectual period, in this respect similar to the Negro. The difficulty only lay in the method that would be used to endow him with speech. I knew of all the fruitless attempts of my predecessors; and it goes without saying that, faced with the competence of some of them and the nullity of all their efforts, my resolution faltered more than once, until all of my deep thought on the subject led me to this conclusion:

The first step consists of developing the monkey's vocal apparatus.

This is, in fact, how one works with deaf-mutes before getting them to enunciate; and hardly had I thought this over, when the analogies between the deaf-mute and the monkey crashed into one another inside my soul.

First of all, his extraordinary mobility in mimicry that compensates for spoken language, demonstrating that just because one ceases to speak does not mean that one ceases to think, though this faculty may diminish due to the paralysis of the other. And then others, more peculiar since they are more specific: diligence at work, loyalty, fortitude, increased beyond a doubt by these two qualities whose commonness is truly revealing: their aptitude for balancing exercises and their resistance to dizziness.

I decided, then, to begin my work with a veritable gymnastics of the monkey's lips and tongue, treating him in this respect like a deaf-mute. As for the rest, his hearing helped me establish direct spoken communication, without the necessity of appealing to his sense of touch. The reader will see that in this matter I was overly optimistic in my expectations.

Happily, the chimpanzee is of all the great monkeys the one with the most mobile lips; and in this particular case, since Yzur had suffered from tonsillitis, he knew how to open his mouth for examination.

The first inspection confirmed, in part, my suspicions. His tongue stayed at the bottom of his mouth, like an inert mass, with no movements other than those related to deglutition. The gymnastics took effect forthwith, as after two months he already knew how to stick his tongue out in mockery. This was the first relationship that he experienced between the movement of his tongue and an idea; a relationship perfectly in accordance with his nature, by the way.

The lips gave us a lot of work, as I had to stretch them out with pincers; but I appreciated—perhaps due to my own expressions—the importance of that anomalous task and carried it out gaily. While I practiced the lip movements that he was supposed to imitate, he remained seated, scratching his croup with his arm bent behind him and winking in doubtful concentration, or smoothing out his side whiskers with the air of a man who harmonizes his ideas through rhythmic gestures. In the end he learned how to move his lips.

But the use of language is a difficult art, as is proven by a child's long stammering, which leads him, parallel to his intellectual development, to the acquisition of the habit. It has been shown, in fact, that the center of the vocal innervations itself is associated with that of speech, in such a way that the normal development of both of them depends on their being practiced in harmony. This was already suspected in 1785 by Heinicke, the inventor of the oral method for teaching deaf-mutes, as a philosophical conclusion. He spoke of a "dynamic concatenation of ideas," a phrase whose profound clarity could teach a lesson to more than one contemporary psychologist.

Yzur found himself, with respect to language, in the same situation as the child who before speaking already understands many words; though he was much more adept at making associations based on the judgments he held concerning things around him, given his more ample experience in life.

These judgments, which must not have been merely impressionistic, but rather inquisitive and disquisitional (judging from the differential character they assumed, which would entail abstract reasoning), gave him a superior level of intelligence, very favorable by the way for my purposes.

If my theories seem too bold, one need only reflect that syllogism, which is to say logic's fundamental argument, is not unknown to the minds of many animals, as if the syllogism had originated as a comparison between two sensations. Otherwise, why is it that the animals that know man run from him, and the ones who have never seen him do not . . . ?

I thus began Yzur's phonetic training.

The idea was to first teach him mechanical speech, in order to progressively lead him toward sensible speech.

Since the monkey has a voice, having in this respect an advantage over the deaf-mute, in addition to certain rudimentary enunciations, the idea was to teach him the modifications thereof, consisting of the phonemes and their articulation, called static or dynamic by the experts, according to whether one refers to vowels or consonants.

Given the monkey's gluttony, and following in this way a method employed by Heinicke with deaf-mutes, I decided to associate each vowel with a treat: *a* with *papa* [potato]; *e* with *leche* [milk]; *i* with *vino* [wine]; *o* with *coco* [coconut]; *u* with *azúcar* [sugar], in such a way that the vowel was contained in the name of the treat, be it the sole vowel sound repeated twice, as in *papa*, *coco*, and *leche*, or the sound reigning over multiple syllables, tonic and prosodic, i.e. as the central sound: *vino*, *azúcar*.

Everything went well as long as we dealt with vowels, which is to say the sounds that are formed with an open mouth. Yzur learned them in fifteen days. Only that sometimes the air held in his cheek pouches gave them a thundery roundness. The *u* was the hardest for him to pronounce.

The consonants proved to be a backbreaking task, and before long I understood that he would never pronounce any of the ones that are formed with the teeth and gums. His long fangs and his cheek pouches got entirely in the way.

His vocabulary was thus reduced to the five vowels; *b, m, g, f, c*, namely all the consonants formed by nothing other than the palate and the tongue.

Even for this task his sense of sound was not enough. I had to resort to touch like with a deaf-mute, placing his hand on my chest and then on his own in order for him to feel the sounds' vibrations.

And three years passed without getting him to form a single word. He tended to call things, as with a proper name, by the letter whose sound predominated in them. That was all.

In the circus he had learned to bark like the dogs, his work mates; and when he saw me lose hope in the face of my vain attempts to wrench words out of him, he barked loudly as if he were showing me everything he knew. He pronounced vowels and consonants on their own, but was unable to associate them. The most he could manage was a dizzying repetition of *pes* and *emes*.

As slow as it may have been in coming, a great change had been wrought in his character. His physiognomy exhibited less mobility, his gaze was deeper, and he adopted meditative poses. He had acquired, for example, the habit of gazing at the stars. His sensitivity developed as well; he began to go to tears with a noticeably growing ease.

The lessons continued with unbreakable tenacity, though with the same results. The whole thing had turned into a painful obsession, and slowly I began to feel inclined to use force. My character began to sour with this failure, until it assumed a deaf animosity toward Yzur. He grew intellectually within his rebellious muteness, and I was becoming convinced that he would never amount to anything greater, when all of a sudden I discovered that he did not speak because he did not want to. The cook, horrified, came to tell me one night that he had caught the monkey unawares "speaking real words." He was, as he told it, curled up against a fig tree in the garden; but his fright kept him from remembering the essential part, namely, the words. He only thought he remembered two: cama [bed] and *pipa* [pipe]. I could have kicked him for his idiocy.

Needless to say, I spent the night possessed by a great emotion; and what I had not done in three years, the mistake that ruined it all, was as much the product of the nervous energy of that sleepless night, as of my excessive curiosity.

Instead of letting the monkey reveal his use of language naturally, I called him in the next day and tried to force it out of him through obedience.

I obtained nothing more than the *pes* and *emes* with which he had tried my patience, the hypocritical winks and—Lord forgive me—a certain glimmer of irony in the restless ubiquity of his grimaces.

I became furious and, without stopping for a moment's consideration, gave him a lashing. All that I won was his weeping and an absolute silence that did not let even a moan escape.

After three days he grew ill, in a sort of sullen dementia complicated by symptoms of meningitis. Leeches, cold showers, purgatives, cutaneous revulsions, bryony alcohol, bromide: all the most frighteningly awful therapies were administered to him. I fought with desperate vigor, driven by remorse and fear. The former for believing the beast to be a victim of my cruelty; the latter for the fate of the secret that he might take with him to the grave.

His condition improved after a long time, though he remained so weak regardless that he could not get out of bed. The nearness of death had ennobled and humanized him. His eyes, full of gratitude, did not leave me for a moment, following me about the room like two revolving balls, even when I was behind him; his hand groped for mine with the intimacy of convalescence. In my great solitude, he quickly began to take on the importance of a person.

The demon of analysis, which is no more a sort of perversity of the spirit, impelled me to renew my experiences regardless. The fact was that he had spoken. It could not be left at that.

I started very slowly, asking for the letters he knew how to pronounce. Nothing! I left him alone for hours, spying on him through a hole in the wall. Nothing! I spoke to him in brief sentences, trying to reach out to his loyalty or his gluttony. Nothing! When all of this turned out pathetically tears welled up in his eyes. When I said to him a customary phrase like, "I am your master," with which I began all of my lessons, or, "you are my monkey," with which I complemented the former statement, to lend his spirit the certainty of an absolute truth, he assented by shutting his eyelids; but he did not make a sound, he did not even move his lips.

He had fallen back on gesticulation as his only means of communicating with me; and this detail, together with his traits in common with deaf-mutes, doubled my worries, since everybody knows of the latter's predisposition to mental illnesses. At times I wished that he would go mad, to see if the delirium might finally break his silence.

His recovery remained static. The same gauntness, the same sadness. It was clear that he had been made ill by both intelligence and pain. His organism had been broken at the hands of an abnormal cerebration: whether he lived six more days or half a dozen, he was a lost cause.

Yet, in spite of the growing tameness that the illness was working in him, his silence, that maddening silence brought on by my own exasperation, would not let up. From down in the dark depths of a tradition fossilized as instinct, his race imposed its age-old muteness on the animal, fortifying itself with an atavistic will at the very roots of his being. The ancient forest men who were forced into silence, which is to say intellectual suicide, by who knows what barbaric injustice, kept their secret made up of the mysteries of the forest and the abyss of prehistory through a decision that is by now unconscious, though formidable in the immensity of its years. The misfortunes of the humanoid who fell behind in an evolution in which humans got ahead through dark and barbaric despotism, surely dethroned the great quadrumanous families from their dominion over the trees of their primitive edens, thinning out their ranks, capturing their females in order to secure enslavement straight from the maternal womb, until they infused the impotence of the vanquished with an act of mortal dignity, which drove them to break that superior but accursed tie of speech with the enemy, finding ultimate salvation in the darkness of animalism.

And what horrors, what stupendously ruthless acts the victors must have committed against the semi-beast in the throes of evolution so that the latter, after having tasted the joys of the intellect, which is the heavenly fruit of the scriptures, would resign himself to that capitulation of his line on the demeaning terms of an inferior; to that regression that forever stunted his intelligence into nothing more than the mechanical gestures of an acrobat; to that great cowardice toward life that would eternally bend, as a marker of his beastliness, the back of he who has been dominated, stamping him with that look of melancholy confusion that lives on in the depths of his caricature.

And this is what, on the very cusp of success, aroused my ill humor in the depths of an atavistic limbo. After all those million years, speech, with its magic spell, stirred the ancient simian soul; but against that temptation that would violate the protective darkness of animalism, the ancestral memory, spread throughout the species under guise of an instinctive terror, opposed era upon era like a wall.

Yzur went into agony without losing consciousness. A sweet agony with his eyes closed, with weak breathing, slow pulse, absolute stillness, which he only interrupted to turn from time to time toward me, with a heart-rending expression of eternity, his face like that of an old sad mulatto. And the last evening, the evening of his death, was when the extraordinary thing happened, the thing that made me decide to take up this narration.

I had dozed off at the head of his bed, defeated by the heat and the stillness at the onset of twilight, when I suddenly felt that I was being grasped around the wrist.

I awoke with a start. The monkey, his eyes opened wide, was finally dying this time, and his expression was so human that it filled me with horror; but his hand, his eyes, drew me to him with such eloquence, that I had to lean toward his face at once; and then, with his last breath, the last breath that both crowned and dashed all my hopes, there issued—I am sure of it—there issued forth in a murmur (how can one explain the tone of a voice that has remained speechless for ten thousand centuries?) these words whose humanity reconciled the two species:

—MASTER, WATER, MASTER, MY MASTER . . . [1]

NOTE

1. TN: The Spanish word for "master," *amo*, can also mean "I love." Therefore the original can also be interpreted as, "I love, water, I love, my master."

III

DARWIN AND THE SHAPING OF ARGENTINE IDENTITY

Historical-Sociological Studies (Bunge, Ramos Mejía, Ingenieros, Sarmiento)

RAMOS MEJÍA: IMMIGRATION AND DARWINISM

It has been pointed out that positivism was the ideology that accompanied and legitimated the economic modernization of Argentina and its state policies, aiding the country's entrance into the world market. During this period, the positivistic state homogenized (bureaucratically and discursively) the diverse social structures (rural migrant workers in the city, politicized labor, delinquents, women, and homosexuals, but most importantly immigrants) in order to make them governable.

Because immigrants' economic successes in some sectors were beginning to displace the criollo class in economic production, the elites attempted to assimilate immigrants into the mainstream progress of the nation, while at the same time holding on to their own hegemonic status. Some of the authors in this section were members of this elite (Ramos Mejía and Bunge, for example), and wrote about immigrants in Darwinian terms in order to explain the sociological phenomenon from a scientific perspective that would also point to their own elite class as the society's most "evolved" or "fittest."

José María Ramos Mejía (1849–1914) carried out a vast body of work as a doctor, politician, and essayist. As stated in the introduction, he was the creator of the Public Healthcare Program, the Department of Sanitation, and the university's Department of Neuropathology, and was a member of the Medical Circle.

In *The Argentine Masses* (1899), an excerpt of which is included in this section, Ramos Mejía takes on the problem of the masses' governability throughout the country's history. For the author, in the gaucho and immigrant multitudes the ability to reason is lost and the individual is "primitivized," returning to prior stages in the evolution of the species. If the masses lose their rational faculties, the issue is how to turn them into citizens, how to bridge the gap between the multitudes' production of wealth for the nation and their ability to become full citizens. This debate is connected with the controversy over universal suffrage and the ruling elite's opposition to it. In 1895 more than 50 percent of male adults were excluded from political society, because the masses were believed to be "incapable of rationally differentiating between the different options on the ballot."[1]

Ramos Mejía goes on to sketch out a descriptive typology of the masses, which he divides up into the *oaf* (*guarango*), the *riffraff* (*canalla*), the *hick* (*huaso*), and the *golden bourgeois* (*burgués aureus*). The last one was the most dangerous because of his "hasty ascension to the Capitol." Ramos Mejía then proposes that public education is the right way to treat and cure the sick social psyche of the immigrant and adapt him to the civic environment.

Ingenieros harshly criticizes *Argentine Masses*. For him, Ramos Mejía's work has great literary value, but lacks scientific rigor. The basic error for Ingenieros is that Ramos Mejía follows Le Bon's *Psychologie des foules*. Neither Le Bon nor Ramos Mejía establish precise limits between collective psychology on the one hand, and the psychology of the masses on the other. For Ingenieros, the former refers to a more fixed and consolidated group, like race, nation, or class, whereas the concept of "the masses" is purely transitory in nature, and thus the term refers to a group with no clear identity. In Ingenieros's view, Ramos Mejía's "masses" are but a hodge-podge of disparate groupings, which proves to be both unproductive and imprecise.[2]

BUNGE: *MESTIZAJE* OR "CROSSBREEDING" IN *OUR AMERICA*

In *Our America* (1903) Carlos O. Bunge (1875–1918) takes this biological positivism to an extreme that anticipates racialist genetics. In the essay included in this anthology, Darwinism is employed to explain the processes of atavism, according to which "*a cross between two distinct varieties of a species reproduces the specific ancestral form.*" In talking about the *ñata* cow, Sarmiento also used this term to describe the cow's apparent degeneration or regression. For Bunge, atavism is the exact opposite of evolution, whereby hybrids and crossbreeds degenerate back to the primitive characteristics of prior species in their lineage, though in the end he does not reach a conclusion as to whether hybridism or crossbreeding between the human races is altogether harmful. However, he does venture the idea that in the human races, "the crossbreed will tend to reproduce a primitive form of man, or at least one that is ancient and pre-Christian." Likewise, Bunge uses the principle of "correlation," a phenomenon studied in the animal world by Darwin himself, to conclude that there is a "correlation" between "crossbreeds'" physical and psychological characteristics, namely "*psychological discord, degenerative semi-sterility, and lack of moral sense.*" Further on he exhibits yet another brand of racism in his long-winded description of the mulatta woman, who for him is the ideal object of desire for the white man.

Bunge goes so far as to bless alcoholism, smallpox, and tuberculosis for having decimated the indigenous and black populations of Buenos Aires, thereby "Europeanizing it." He also advocates for the use of anthropological methods to watch for and detect how assimilated "crossbreeds" cover up their hidden traits.

Ingenieros points out that Bunge is mistaken in placing so much importance on the correlation between physical and psychological characteristics. Race alone is not a determining factor, since one must also consider the natural and economic environment in which race exists. For Ingenieros, as the reader will see in this section, economic factors constitute the main driving force that guides social evolution.[3]

Along the same lines as Bunge, a part of Sarmiento's essay, *Racial Conflict and Harmony in the Americas* (*Conflicto y armonías de razas en América*) has been included in this section. As José Ingenieros saw it, though Sarmiento and Bunge both attempted to devise a scientific explanation of race in South America, only the former was successful.

INGENIEROS: THE NATURAL EVOLUTION OF IMPERIALISM

José Ingenieros (1877–1925) carried out his research in psychiatry and criminology as the clinical director at the Buenos Aires Police Department's Observation of Outcasts Service, and at the national penitentiary's Institute of Criminology. This author differs from the other authors in this section in several ways. As we have pointed out, he rigorously criticizes them, and his thorough understanding of Darwinist theory certainly gives him the background to do so. Notable about his vision is that it combines aspects of positivism and Darwinism with socialist doctrines and economic theory. Ingenieros understood sociology as a *natural science*, and set out to test a genetic interpretation of Argentina's evolution, based on the premise that all sociological phenomena are merely the combination of a series of physical, biological, and psychological circumstances. Man, for Ingenieros, "is not an aerolite fallen to earth through some whim of the supernatural forces," but rather "a living creature, nothing more. . . . He fights for his life in order to satisfy basic and indispensable needs: the preservation of the individual and the reproduction of the species." As opposed to Ramos Mejía and Bunge—both followers of Spencer—Ingenieros thought that humanity, as a biological group, did not have a mission in the universe and lacked absolute moral principles. Thus, for him, evolution lacked any deterministic goal of progress. For Ingenieros, "[t]he spring that sets man's social activity—his conduct—in motion is the sum of his needs; his consciousness of the latter—subjected to rigorous determinism—is merely the apparent motive behind all individual or collective actions." Thus, Ingenieros believed that biological sociology should replace Spencerian organicism.

Ingenieros employed this biological perspective on social evolution to depict imperialism as the leadership of the fittest in an international struggle for survival. Ingenieros saw expansionism as a phenomenon which was subject to natural and scientific laws outside the realm of moral judgment, and, because of his confidence in the elite leadership of the "fittest," he predicted that Argentina would become the supreme political power in South America, one which would be able to oppose North America's United States on equal footing.

SARMIENTO: A COMPARATIVE PERSPECTIVE ON IMPERIALISM

Sarmiento's *Racial Conflict and Harmony in the Americas* (1889) is an impressive mass of bibliographic research which attempts to scientifically explain the history of the New World's inhabitants in terms of race, biology, and psychology. In this book, as well as *Facundo* and others, the author strives to analyze the similarities and differences between the histories of North and South America (Sarmiento's admiration of the United States of America is well documented). Sarmiento's ideas in this vein are thus similar to those of

Ingenieros, in that both authors foresee an imperialist future for Argentina, a goal which they both believe can be achieved through the naturally evolving path of history, and by following the United States' blueprint for progress.

It is interesting to note that both Sarmiento's *Racial Conflict and Harmony in the Americas* and Alberdi's *Luz del Día's Pilgrimage* were republished in 1916 in José Ingenieros's collection titled *Argentine Culture* (*La cultura argentina*), which indicates that there was still an ongoing debate in the intellectual field concerning the texts included in part II of this book. Ingenieros thought that Sarmiento's work was a foundational text for the scientific study of sociology in Argentina, and decided to republish *Conflict and Harmony* before republishing the canonical *Facundo: Civilization and Barbarism* (1845).[4] It is worth pointing out that it was after writing *Facundo* and before *Conflict and Harmony* that Sarmiento first read Darwin and Spencer, and incorporated their ideas into his own explanation of how South America's population had evolved.

Among all the essays collected in this anthology, Sarmiento's *Conflict and Harmony* is the only one that addresses the inclusion of the African American population in the cultural evolution of South American countries. Sarmiento explores the topic by quoting from studies of the situation of African Americans in the United States (Prescott and Wilson), which gives him pertinent material for a comparison with slavery in other parts of the Americas. This last essay culminates in a contradictory mixing of racism and philanthropy, in which the author predicts that the African race is on the verge of a spectacular resurgence.

NOTES

1. Oscar Terán, *Positivismo y nación en la Argentina* (Buenos Aires: Puntosur, 1987), 22.
2. José Ingenieros, *Sociología argentina* [Argentine Sociology] (Madrid: Daniel Jova, 1913), 131.
3. Ibid., 202.
4. See Fernando Degiovanni, *Los textos de la patria* [The Texts of Our Homeland] (Rosario: Beatriz Viterbo editora, 2007), 277.

THE ARGENTINE MASSES (1899)

José María Ramos Mejía

THE MULTITUDE IN MODERN TIMES

[. . .]

As I was studying the amazing method that nature has devised for the development of living things—all the way from our humble Silurian beginnings, the primitive fish, up to man—it seemed to me, and rightfully so, that an analogous process must have taken place in the formation of our society. Upon reaching a certain stage in its development, the initial embryo, the immigrant, must have adopted something socially analogous to the anatomical structure of a fish, and then later on that of an amphibian, and finally that of a mammal. By this I mean to say that he must have undergone a similar sort of transformism in the successive steps of his intellectual and moral refinement. In reality assimilation must have been quite enticing for him, because it provided his spirit with a comfortable and attractive method of inquiry. Moreover, in what we can rightly call our social phylogeny he came across different types of beings that, with a little good will he could take to resemble himself, in a far-off but enlightening comparison.

Any *numbskull* is smarter than an immigrant who has just arrived on our shores. He is somewhat amorphous, *cellular* even, in terms of his total lack of an even moderate progress in the organization of the mind. He has a slow brain, like that of the ox he has lived alongside: myopic in terms of mental wit, clumsy and dull of hearing in everything having to do with the spontaneous and easy acquisition of images by way of the brain's great sense. Could there be a darker faculty of perception? Could there be a more cumbersome physiological conductor, one which was any clumsier at transmitting even the most elementary of sensations, than their pachyderm-like skin?

I remember that, years ago, on account of an epidemic in the "Immigrant Shelter" and under the pretext of investigating *mysterious states of immunity* in the healthy and newly-arrived residents, I carried out some psychological experiments, that while not entirely conclusive, as is logical, allowed me to make approximate deductions, as they afforded me curious and revealing results in this respect.

All of these easy methods based on fairly uncomplicated argumentation permit even the most ignorant of people to question, through material means and with relative success, that dethroned *sphinx* known as the spirit's impenetrability. Simple methods like *the smallest perceptible differences* or *true or false scenarios*, as with any other research procedure, afford this sort of interesting results. Ordinarily any of these simple conscious impressions will provoke a reaction in the subject, one that is not merely motor, but intellectual as well; the attention is drawn to the sensation, judgment steps in to classify it or define its characteristics, and an emotional state accompanies the whole unit. This is how nearly the entire spirit comes together to participate in the simple production of an elementary sensation.

Crepuscular, then, and larval in a certain sense, is this peasant's state of psychological advancement (a peasant who is, in part, the vigorous protoplasm of the new race) as he sets foot on our soil. It is important to understand that this dull boor really does not *feel* the way we do. As I said earlier, his psychological apparatus is slow and intermittent like some primitive spinner's wheel, or the rough plow of a Medieval farmer. Moral sensitivity, which for a fully-formed and refined man is the receptacle for and factory of all moral feelings and ideas, is still a vague parody of what it will go on to be. But the *environment* works wonders on the moldable tameness of his nearly virgin brain. The light of this sky awakens his dormant use of visual imagery. The noise first and then the sound, the varied color, the many shapes of things, and that secret inclination and elemental ability of this race in art—not in the loftier sense, but in some of its more humble (though by no means less useful) manifestations, namely the manual and domestic arts that put bread on the table and earn a living—come together to this end. He is awakened by the vigorous locomotive that huffs as it pulls along its lengthy tail with its rings of unending wagons, choking on the yields of the generous harvest. He is awakened by the noise of the streets, the bustle of industry, the happy cries of the children that *sprout up* on the patios of the little convents like corn in the lush soil. Finally, the immense plains, those unrivalled plains of ours, without a shadow in sight, with its far-off, melancholy horizons, covered in wheat fields and green cornfields. It is all as he had never imagined it, not even in dreams of delirious grandeur, that *bumpkin* who is so fertile under this sun, in this air, upon this immense river of our fatherland, gently roaring along its lengthy course.

[. . .]

I am surprised by the docile moldability of this immigrant Italian. He arrives on these shores amorphous and protoplasmic, and accepts with prophetic tameness all the shapes impressed upon him by both want and legitimate ambition. He is anything and everything in the life of the cities and the countryside, from roving musician to cleric. With the same hand that gives a blessing—full of that comical solemnity of one who does such a thing as a trade and not a calling—he turns the crank of the barrel organ, or pulls along the produce cart. He offers us cheap umbrellas when it starts to rain, he makes the nimble monkey dance atop a tripod, and he opens up the earth that he has conquered with his perseverance and made fertile with his work. There are so many of them that they flood everything: the second and third class theaters, the free promenades, the churches (because they are devout and tame believers), the streets, the squares, the shelters, the hospitals, the circuses, and the markets. They take on as many trades and professions as their slightly backward and elementary attitudes will permit. They are coachmen, after a painful learning process, full of bumpy rides and crashes into other buggies, of tickets and violations, all of which goes to sharpen their wits. They are *tow trucks* for the streetcars, striking extravagant and picturesque poses as they wield their reins. They are stage drivers and conductors, and one can even hear saucy urban street slang spouting from their lips, with a certain exotic grace for such a coarse and watery brain. They are, in short, everything that might provide a means of living and the promise of a future, one that though far off, is certain all the same.

Suffice it to say that there are certain jobs from which they have ousted even the *gaucho*. When you go out a ways, a strange sort of burlesque centaur irks your eyes: on top of a hairy and ill-groomed charger, with a twitching tail and a robust gallop, there rocks to and fro a nervous figure whose legs shake to the disjointed rhythm of the nag, which has been battered by exhaustion. As he passes by the general store they whistle at him and shout insults. His anti-aesthetic figure arouses hilarity, but be carries on towards his destiny: he does not accept the *drink*, or the *nightcap*, or the *shot*, or the *glass*, or the *gargle*. He goes on to his purpose: he collects the money for the bundles of brushwood he has sold, for the cowbell he has adjusted, for the mending of a pot, for the barrel organ he lent out, or for the whitewashing of a house, and he returns straightaway to the *stand* or dwelling where he has wasted no time in hiding away his savings in some safe spot, the succulent product of his unceasing work and his *Fregoliform* multiplicity of humble though advantageous abilities. Because, in effect, this unpleasant fellow from Fregoli is none other than the living symbol of the Italian immigrant. With his Basque or runaway dress, his colt skin boots merrily matched up with a frock coat, wearing espadrilles or elastic leggings and a *chiripá*, in the borderlands or on the outskirts of town, in the neighborhood or on the distant ranch, where the struggle against nature's indifference never ends, he—always tame, happy but discreet, tolerant, and extremely docile—gains ground against the circumstances that surround him, and unreluctantly starts to look a lot like the things that the earth places before him and the races that surround him. And so, now and again you'll see as the loyal and faithful husband of one of our women, the lover of a negress, or the devoted beau of a succulent Indian girl, or of some centenarian old maid, whose stagnant *little capital* he makes grow to surprising proportions through the honorable alchemy of his labor, which is full of surprises and transformations.

Even his outlandish innocence at carnival time when he dons such strange costumes, of *Indians, dukes, gauchos,* or *warriors*, has something kind and brave about it. It reveals the drive of new blood, a certain good-natured skepticism that scoffs at the ridiculous man in the street, with his chit-chat and childish bravery. You can see him cross a street that is packed full of people just waiting for a chance to *mock someone*, and there he goes laying down his espadrille-clad pachyderm feet with terrible aplomb to the rhythm of his clogged accordion. He travels all throughout the city, without interrupting his sincerely jovial pilgrimage of three days and their respective nights for even a moment. He enjoys himself like a child, because he is one. Though in years he is an adult, his spirit has only begun to live once his wings, with the loud flapping of a bird that receives the cool blessing of rainwater on a summer evening, feel the fiery and stimulating influence of this light and this sky, which smell of freedom and of hard work. The gargantuan mirth of this foreigner making the best of life is simply contagious! Look how he soaks up his broth at dinner time in big, noisy, appetizing slurps, and leaves behind nothing, not a single embarrassing scrap, in a steaming pot that is full of that health which only noble poverty can bring! This lively *paise* strikes me as such a good and simple person, before he leaves behind the immigrant larva in order to become an unbearable and voracious *golden bourgeois*!

A child conceived in this full effervescence of the nervous system inherits a greater array of aptitudes than if the parents had remained in their initial inert state. For all the reasons here enumerated, the transformed immigrant neither thinks nor feels with his imported instrument, which was deficient, but rather with the one that was poured into the mold imposed upon him by the environment; and so the hereditary influences that are transmitted must be thirty-five percent indigenous Argentinean.

In this struggle in which modern Argentinean society takes its shape, the capital of hereditary acquisitions that he carries with him is necessarily modified by selection. The process of development that was passed to him by his forbears—what Lang would have called palingenetic inheritance—is noticeably modified since this inheritance is easily modified in the struggle for existence. For Vandervelde and Massart this is an important factor, which modifies the flow of the successive states in a species' development on a variety of different levels, such that the course of the offspring's development quickly ceases to faithfully reproduce that of their forbears.

In our country, which is in the midst of its formative enterprise, the first generation following the immigrant, who is the most genuine child of its environment, becomes, in a way that is still ill-defined, the depository for future sentiments of nationality, according to its modern conception, naturally.

If you observe him in his smallest actions and in the various manifestations of this sentiment, you will see the first outlines, however childish they may be, of what is to become the passion of the future, especially insofar as external and sensory worship of the *homeland* is concerned. The little rogue, the child who has been part-way *Argentineanized* by his environment and his heritage, is the vector of that affection that has been present since the time of his birth. The father, too, only now begins to truly feel this new sentiment, for the prosperity of agriculture and our docile, generous soil have been whispering things in his ear. It is as if the scent of the moist earth, on the verge of being fertilized, aroused in the peasant the vague notion of patriotism.

This curious and wandering child who is always on the street, applauds louder than anyone else for the cadet schools that parade with their endearing seriousness on national holidays. He is the one who experiences the tattered flag of the old and glorious battalion with wild enthusiasm, the one who follows the ranks the farthest, the one who is never missing from the list, the one who with the most honest and sincere determination signs up for any and all low-class activities involving the national colors and a uniform.

This initial product of *immigration*, the Argentinean of the future, spends more time in the street that in any other city in the world where childhood is more or less disciplined. The child, as soon as he has been weaned, does not leave the doorway and the sidewalk, the rights to which he disputes with the passersby, and once he can move about on his own he can be found in the square and at the in doors to shows and to the State's abundant schools, roving and wandering about in his merry vagrancy. Here his nervous system is daily receiving and processing a greater number of impressions, and he is the one who most quickly and most intensely finds out about even the most minor public incidents. As a consequence, his brain has been more stimulated and whipped into shape. Since a child's brain can only take in *what it can get*—a fact which outweighs the slightly exaggerated fear that they are overworked at school—he is more precocious in his development than children from well-to-do homes, than the *rich kid*, as the society's slang would say. This probably explains why he is better than the rest in his schoolwork, and why he develops patriotic sentiments with such ease, an element which in the following generation will be even more complete.

Systematically and with compelling insistence they are told about their homeland, about the flag, about the nation's glories and about history's heroic episodes. They hear the anthem and sing it and recite it with the comical furrowed brow and fervor of epic poetry. They comment upon it in their own style with enchanting naïveté, and with their loquacity in action they demonstrate just how right their age is for the sowing of such a noble sentiment. I have always adored the motley hordes of poor children, who, when the hour comes, gush out of the public schools in merry and copious streams, like water out of the mouth of a pipe that has suddenly been opened, flooding the street and filling the neighborhood with their charming din. These stray birds, with such unkempt plumage and with their funny and exotic first and last names, come from an unfamiliar nest, no doubt, but like the mysterious *Rodostetia rosea* that Nansen came upon in his path, they tend to be more successful at flying high and resisting the crude temperatures around them.

The first generation is often deformed and less than beautiful up to a certain age. They look like the product of some crude mold, preliminary castings with a noble metal, but still full of lumps and angles that later polishing will set right. There is a fair share of pug noses, large ears, and thick lips, as their morphology has yet to be modified by the chisel of culture. In the second generation, one begins to see the corrections imprinted by a civilized and more cultured life than the one provided by the immigrant peasant. The change in nutrition, the influence of the air, and the relative peace of mind resulting from the ease with which sustenance and life's primary needs are taken care of, all work a transcendental influence on him.

[. . .]

One must observe the children of the final ranks, in order to see how from generation to generation the immigrant's form is modified and *made into a person.* We could tell, in the presence of any of the many primary school classes, which of them belongs to the first generation, which to the second. We could tell which of them have cultivated parents and noble lineages, fine-tuned by good living, or by the fitness of a body disciplined through constant correction. We could tell which of them have received the constant guidance, the serene and easy practice of moral duty of an impeccable father, or the good example, the constant sound of the maternal voice, full of that musical unction emitted by an organ that has not been coarsened by blasphemy or wild shouting, over the lost ox, the spilt pot, or the wasted shoemaker's wax. In short, which have received all the things that little by little fill the soul with the full-bodied perfume of ideals and ambitions that are *chères au coeur, que mon esprit rêvait* and that find refuge, with a certain masculine modesty, in the humble shade of the ancestral home, that perennial emblem and clear ideal of a life free from the brutal obsession with making one's fortune *at all costs.*

From the immigrant, haphazardly modified thus, there arise, as if out of social epigenesis, all of those products of evolution with which we rub shoulders on a daily basis and that form one peculiar and complete structure.

In the social world, and here we turn back to our the intriguing comparison we made at the outset, the same thing occurs as in the rest of nature, whose harmony requires that the entire fauna of a region contain, apart from the great quadrupeds, creatures of smaller strength and size.

Side by side with lions and elephants there are smaller animals that live off of what they leave behind, and that in compensation for their weakness have received faculties that give them access to things out of the large mammals' reach. They are members of the *small fauna*, which have been described in an admirably comprehensive style by the author of *Les enchaînments du monde animal*, and which have their counterparts in the moral world.

In social paleontology, the *oaf* represents one of those vertebrates that will be searched out with curiosity by the sociologists of the future, in order to establish the succession of biological forms in our evolution. His queer taste is the opposite of art, and he resembles those queers in the sexual sense, who reveal their dubious mental state through the atrabilious manifestation of their appetites. He needs vivid colors, shrill music, like the erotomaniac needs the intense smell of the flesh. He likes gaudy and tasteless combinations of things, like the aforementioned with their twisted attitudes and their rough practices, in order to satisfy the special idiosyncrasies of his senses.

In music, he exhibits the atavisms of the barrel organ that his parents used in their poverty. As far as painting is concerned, his retina holds on to the loud colors of the oleography that surprised his parents when they first came into contact with the affected artwork of the outlying neighborhoods and border towns. That eager but still narrow brain gives social birth to all the visual, auditory, and moral sensations that his predecessors had gestating inside of them: from the sounds of the deep dark green cornfield, the harsh smells of the unthreshed heaps of grain, to the nightly rumination of their tyrannical savings plans, which were really a kind of shortage but were *lived out* like destitution. All of this mixed up in topsy-turvy combinations is what will later show up in his baroque clothing, in how he decorates his home, with his excessive use of *notions*, in his literary tastes, in his innocent parties, in his florid signature, in those *sui generis* perfumes of his and, finally, down to his funeral, full of compromising little urges to giggle brought on by the characteristic over-abundance of swarthy brunettes in mourning garb, lugubrious tapestries, hats, and lusting black horses, who, romping about with great solemnity, trot down the street chasing after any equine animal whose sex is in doubt.

The ideas that an *oaf* has about taste and about art could only be shared by a queer. True it is that the latter is sick, and the former primitive, an innocent exhibitionist; perhaps his chromatic sense in sight and sound has atrophied, giving rise to that poverty of taste and opportunity in all things that life has to offer. The *oaf* is, then, a *stratum* in the special geology of our society.

He has been blessed with the university's habitual injections of *instruction*, but he is a *cultural beggar*. His nearness to the rustic, his next of kin, has made his bark so rough that it resists the *vernissage* which the traditional home should provide, or, in its stead, the culture of the university when the university is not just a professional, utilitarian affair, as is the case with ours. And so, even when you see him working as a doctor, lawyer, engineer, or journalist, from a league away you will pick up the pungent odor of the stable and of the quarters of an *oaf* in full, from head to toe. You will see him find his way into the finest society, become a member of the finest clubs, and masterfully resist using a brush. You will see him make an effort to mend his ways and he will mend them, sometimes; but when you least expect it, that stubborn constitution of his will shine through, because it will take a couple more generations before he can leave behind the larva that still clings to the initial state.

They are driven by their sensibilities, though in some cases with a certain modest discretion, to secretly love the comical lyricism of *Flor de un día*, which is the *canon* of their literary tastes: to cry over Lola's emotions and interesting don Diego's misfortunes. They still live in that period of mental development in which one can honestly admire the *Catalan* Serrallonga, and love speckled carnations and the pompous, garish flowers that even as long as thirty years ago filled the air of our brick patios with their kind and penetrating smell. As for lovers' suicide, that lovely example of a social regression back to the Wertherian period of the novelistic *lummox*—could you find it anywhere but in the innocent *oafling* couple, whose emotional innervations, knocked out of proper balance, have taken hold of their judgment? That anachronistic self-mutilation is a painful demonstration of the influence that such touching books as *Oscar and Amanda, Love or die*, and *Pablo and Virginia* still hold over the primitive heart. The *oaf* is such a transitional social type that it actually disappears in the third generation, making way for a new type.

The *riffraff* is an *oaf* who has climbed up the ladder of fine dress or of money, but whose soul is still full of atavisms, in whom we could say, making use of the *sui generis* anatomy of laymen, that the liver, through some anomaly of this species, spills its venom onto the heart instead of sending it down the intestine. Throughout the beast's long process of domestication, there is something that eludes the works of time and of instruction, something that remains in his soul permanently, like a mole that endures on the skin despite the constant renovation of its epithelium. It is something that, in imitation of the coarseness of the Haekelian system of classification, I will call the *riffraffian appendix*. Like those organs that live on in man's body in a rudimentary state, as reminders of his zoological past, such as the mysterious pineal gland, that harkens back to the third eye of the distant batrachian, or like the caudal appendix, which, though atrophied and ridiculous, can still be found on the human spinal column, reminding man of his embarrassing simian ancestry, so too do there endure in certain men moral habits and behaviors that immediately reveal the riffraffian soul that life has dealt them. Scratch away at the varnish with which social benevolence has deceitfully coated the cracks in that *anima maculata* and you will see the appendix in question for what it is, nullifying the amnesty conferred upon him by the tailor and the impunity of some pompous diploma.

The *hick* is a more grotesque species of oaf. He has true pachyderm skin that scratches one with its coarse touch, and from which, through the workings of the environment, comes the *chum*. The *chum* is a *hick* at heart, but has been tempered by contact with city life and by constant subjection to work in the street, which puts him in daily contact will all the social classes.

There is another variety of *oaf* that differs from the others in the lower degree of exhibitionism in his life and tastes, which are sparse and highly conservative. He also has a certain shrewd modesty, which is the result of his greed, and perhaps of the fear that the great fortune he has amassed at the cost of his health might be discovered. In our land he plays the role that the bourgeois plays in others, the makeshift millionaire born through the magic of the lottery, who sprouted in the immense fields of the backcountry, or from the steaming mound of earth fertilized by his own noble work. But once he has taken up his position, he has no greater plans for life than to save his money, defend it from the charity and patriotism that come knocking at his door from time to time, press it against his chest so that it will stay flat, water it with the milk of *sale on reversion* and of *mortgage* so that it will multiply, clasped against the teats of the usury that keeps fortunes safe from consumption, and is a comforting balsam for ruin and pain. Once he arrives at these heights, he buys, with little money, naturally, a title: it is sometimes called *High Commerce*, for example. Because his plan is the one noted above, you will see him, spruce and solemn, tamely following behind Mr. Juan Manuel de Rosas, admiring García Moreno, or smiling at Santos and Melgarejo, without putting too much thought into matters of good governance. They are souls without a care for any ideals that don't produce money, and they think the same of despotism as of liberty, as long as it leaves their money alone. But . . . woe to the government suspected of disrespecting another man's penny! Because then the man made rich will rise up in heroic revolution and give up his life . . . before giving up his money. In this sort of passive role he is the receptacle and even the incubator of all the political and social gossip sold as journalistic literature by the multifarious press. Since he is as credulous and gullible as *Sancho Panza*, one need only write something unctuous and easy-to-digest, and start it off with the amusing solemnity that our reporters use to grease things and make even the most unlikely news or libel spread. They say, *to the honorable gentlemen, to the sensible sector of public opinion, to the enlightened populace*, and he, not even suspecting that the one writing the article is likely neither honorable, nor sensible, nor enlightened, takes the bait and swallows the hook, going so far as to digest it, such is the strength of his digestive juices.

This *golden bourgeois*, in large numbers, will be formidable if our national education system does not modify him with the scrub brush of culture and the percolation of other ideals that might hold him back from his hasty ascension to the Capitol.

OUR AMERICA

(1903)

Carlos O. Bunge

BIOLOGICAL PRINCIPLES OF CROSSBREEDING[1] AND HYBRIDIZATION

Crossbreed is the name given to the offspring of two animals belonging to distinct *varieties* of the same *species*; a *hybrid* is the product of individuals of two different *species* of the same *genus*. A foal born of an Arabian mare and an Anglo-Norman colt is a crossbreed; and a hybrid would be the mule resulting from that mare and an ass (since the horse genus comprises the horse itself, the ass, the zebra, and the quagga). As a general principle, biology teaches that crossbreeding does not harm the offspring's vitality while hybridization does. Yet, as it is still difficult to readily distinguish between a variety and a species, and since there are varieties that tend toward species status and vice versa, we hold that in many cases crossbreeding and hybridization are confused with one another. The fact is that, in reality, nature does not neatly separate the variety from the species, given that all varieties are inclined to become species and all species originate from primitive varieties. As such, hybridization is crossbreeding in a broad sense; and crossbreeding, hybridization in a restricted sense. These concepts, more than distinct phenomena, represent various gradations of one single phenomenon.

In the human genus the problem is even more complex. Is hybridization possible? Are the four races—White, Yellow, Black, and American—species or simple varieties? At the very least, if they are to be considered varieties, they are varieties with an inclination toward species status; if they are species, they are species with variety-like tendencies. Herein lies the difficulty in asserting as a general and absolute rule whether or not human crossbreeding is favorable or unfavorable for the offspring's vitality.

To narrow down the question, what happens when we cross White and Black, or Black and Yellow, or Yellow and American? Although in this case experimentation is not possible, since we cannot breed men like dogs, observation and experience can provide us with some data . . . And, above all, biology itself. Because biology furnishes us with two or three general principles that are applicable to all breeds, varieties, and species: *atavism* in the crossbreed, and the *correlation* between somatic and psychological traits.

Atavism. Darwin observes that, if one crossbreeds several pigeons belonging to different varieties, the offspring lose these varieties' distinctive characteristics and tend to reproduce the ancestral form of the wild species *Columba livia*, of which they are all descendants. The *ponter* loses its enormous crop; the *runt* its thick beak and its large wings and feet; the *barb* its short and wide beak; the *colipava* its large tail feathers; the *turbitt* its short conical beak and the inverted feathers on its breast. The many and varied crossbreeds of all of these varieties will always tend toward the atavistic form of the *Columba livia*, with grey wings striped with black.

In these cases we are dealing with "aberrant" traits that the breeder has produced in distinct varieties of the same species. But an analogous atavism can be observed in the hybrids of distinct natural species. Crossing different breeds of horse, and in particular horses and asses, one observes that the crossbred or hybrid offspring frequently exhibit a darker dorsal band and striped skin around the legs. Out of ten mules, at least nine display this zebra-like trait, which, when a horse is crossbred with a zebra or a quagga, is undoubtedly stronger and more marked. This led Darwin to infer the existence of a primitive horse that was naturally striped, from which the four or five branches of the genus descended, and also to propose the clear

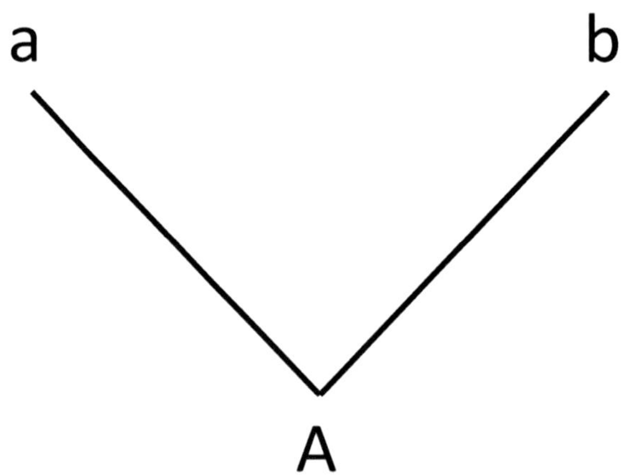

Figure 1.

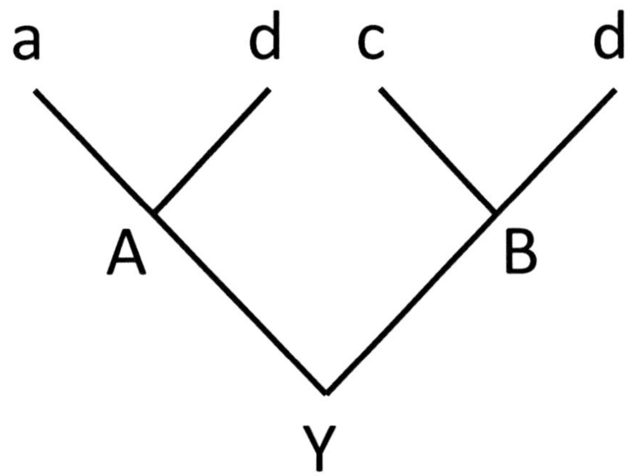

Figure 2.

principle that *a cross between two distinct varieties of a species reproduces the specific ancestral form; between two species of the same genus, the generic form.*

In the case of simple crossbreeding, the principle can be represented by the diagram in figure 1.

The varieties *a* and *b* descend from the specific form A; ergo, crossing them, one can obtain the following atavistic form: $a \times b = A$.

In the case of hybridization, the diagram is more complex (fig. 2).

The generic form *Y* produces the species *A* and *B*, which in turn engender the varieties *a* and *b*, and *c* and *d*. If the varieties *a* and *b* are crossed, the result is a crossbreed with the specific traits of *A*; if the varieties *a* (species *A*) and *c* (species B) are crossed, the result will be a hybrid with the generic traits of *Y*: $a \times c = Y$.

If we apply this criterion to the human races, the result is that the crossbreed will tend to reproduce a primitive form of man, or at least one that is ancient and pre-Christian. Naturally this is a very vague and general tendency, full, as we shall see, of palliatives and logical exceptions . . . Such is the principle of *atavism* in human crossbreeding.

Correlation. Perhaps a simpler principle is that of *correlation*, which could be formulated thus: there is not only a necessary correlation between the physical and psychological orders; there are also *physical and psychological traits that are apparently independent of one another, but which are in fact interconnected and complementary.*

Darwin notes that "colour and constitutional peculiarities go together, of which many remarkable cases could be given among animals and plants. From facts collected by Heusinger, it appears that white sheep and pigs are injured by certain plants, while dark-coloured individuals escape . . . ."[2] Wyman observed in Virginia that all the pigs were black; when he investigated the cause, he found out that these animals eat the root of a plant called *paint root* (*Lachnanthes*) which, in the case of the ones that are not black, turns their bones pink and makes them lose their hooves. Thus, since only the black ones were likely to survive, the farmers killed all the rest of them at birth. "Hairless dogs have imperfect teeth; long-haired and coarse-haired animals are apt to have, as is asserted, long or many horns; pigeons with feathered feet have skin between their outer toes; pigeons with short beaks have small feet, and those with long beaks large feet. Hence if man goes on selecting, and thus augmenting, any peculiarity, he will almost certainly modify unintentionally other parts of the structure, owing to the mysterious laws of correlation."[3] All of these observations prove the correlation of distinct *physiological* traits between one another.

Psychophysics, in turn, demonstrates the correlation between *physical* and *mental* elements. And, as far as the correlation between different *psychological* qualities is concerned, for lack of another text, one could use as evidence my long exposition in Book I about the psychology of Spaniards, in which we see that all of their most characteristic traits concur in one central point that I have called "arrogance." Thus, the principle of correlation is applicable in three orders of biology: the physical, psychophysical, and psychological realms.

This principle, when applied to human crossbreeding and especially to the criollo,[4] gives us, as a direct consequence, the following: the criollo's physical form has a corresponding psychological form. Moreover, any criollo's level of indigenous or mulatto spirit has corresponding anthropological traits characteristic of crossbreeding. If, for example, in a family ten offspring out of which nine are born with a European physical form and one with a Negroid or mulatto form, the former possess a European psychology; the latter, a mulatto psychology . . . In a word, any physical crossbreed, whoever his parents and siblings may be, is a mental crossbreed.

[. . .]

GENERIC TRAITS OF SPANISH AMERICAN CROSSBREEDS: PSYCHOLOGICAL DISCORD, DEGENERATIVE SEMI-STERILITY, AND LACK OF MORAL SENSE

As generic traits of all Spanish American crossbreeds, be they Hispano-Negroid, Hispano-Indian, or Negroid-Indian, I will cite these three: a certain *psychological discord*, relative *sterility*, and a lack of *moral sense*.

One frequently observes in hybrid animals, the children of parents from distinct species, such as the mule, a curious duality of ancestral traits, a double and alternating nature, which tends sometimes towards one parentage, and others towards the other. In the dog-wolf crossbreed, for example, we find a conflict between and alternation of the wolf's fierceness and the dog's loyalty. The same can be said of human hybrids, if they indeed exist; they have not one head, but rather two halves of a head . . .

This dualism that I call "psychological discord" becomes truly unpleasant and tragic in the age of impulses, the youth of able crossbreeds who, conscious of their personhood, live as though they were constantly surprised at themselves! Irritability is hyperactive in their psychology, not just because of their lack of years, but also because of the leaps they must take. Mulattoes' springtime is black! However, their years of maturity tend to be white . . . Because only then, after so many internal struggles, they instinctively find a balance, a *modus vivendi*, which gives primacy to either of their tendencies, to the strongest one or to the one best adapted to the environment. One could say that they nurse a calloused body that serves as a bridge between their two half brains, which welds them together and . . . they calm down. But not before a number of savage brawls, Spanish-style insurrections, and criollo-style revolutions.

In Indian crossbreeds, who are by nature weaker and quieter, the duality either whittles them down into misery or makes them hysterical to the point of heroism. Sadly, many, many more of them fall into misery than those who border on heroism . . .

This psychological discord can be seen as a spiritual form of anthropological hybridism; and this, in turn, as a zoological form of what in botany has been called *disjunction of hereditary traits*. What naturalists usually apply to this "disjunction" is the illuminating, though not infallible, "Law of Mendel." The phenomenon is extremely complex. It essentially consists of the *alternating production*, in hybridizations of plants and some animals, of *disjunctive forms and mixed forms*. Sterility is generally greater in the "mixed" ones than in the "disjunctive" ones.

In animal species hybridization produces some types of sterility that naturalists, who still follow Brocca's classic nomenclature, classify thus: *agenesis*, hybrids that are sterile from the first generation; *disgenesis*, hybrids that are sometimes fertile in the first generation, but are totally sterile in the second; *paragenesis*, or *collateral hybridity*, first-generation hybrids that are only slightly fertile, and second-generation hybrids that, through regression, are fertile amongst each other; and *eugenesis*, hybrids that are directly and indefinitely fertile from the first generation on.

And here it is fitting to ask once more: Can there be hybridization in the human genus? Are the "human species," in relation to one another, true zoological species? . . . The only possible answer seems to me to be the one expressed earlier: the notion of species is relative, and, as such, is applicable to the human genus. The polar bear, with its white fur and long, flat skull, and the black bear, with its oval-shaped brow, long considered by naturalists to be different species, have now turned out to have one close and common origin, and may be equally or more related than the many human races. And what are we to say of the silly lap dog and the powerful St. Bernard? . . .

From the outset, we can venture that the anthropological traits of the yellow race so separate it from the white, that they could very well have descended from two distinct varieties of the *Anthropopithecus*, a distant ancestor of man. On the other hand, the black and white races seem to be more closely related, either due to a less-distant common origin, or to contact and crossbreeding during the historical period.

In short, peoples' centuries-long migrations and the consequent mixing have created so many affinities between the races that, if there is hybridization, it is complex indeed . . . This is why we can only analogically apply Brocca's nomenclature regarding sterility to the human "species." Also by analogy, it could turn out that the crossbreedings of the human species in Spanish America generally resulted in *eugenesis* and in *paragenesis* only as an exception; their *sterility*, if it were to exist, would be, then, very moderate. Save in the case of certain odd mixes of Indian, Black, and Chinese—authentic Chinese, from Imperial China[5]—produced in the nations of the Pacific, especially in Peru, the crossbreeds of America are prolific, and often extremely prolific . . . We should believe this because the statistics tell us so, and under no circumstances—as some malicious and Sancho-Panza-like souls might suppose—because of the saying, "weeds never die." Who believes in these folk saying anyhow, and who dares call the likes of Santa Cruz and Porfirio Díaz weeds!

However, though Spanish American crossbreeds may be fertile, they tend towards a more pure or defined form: European-like in cold climates, Indian-like in warm ones. Hybridity in and of itself is not maintained; it constitutes a passing state, one that is absolutely transitional. If it is repeated and perpetuated through continual crosses between similar crossbreeds, the line degenerates, until there arises what psychiatrists call "the dissolution of the species through degeneration," meaning they engender forms that are so abnormal as to result nearly or wholly inadequate for reproduction. It is true that degenerate individuals are born in even the most pure or well-defined human species, due to mysterious causes that the ancients believed were the rage or inspiration of their gods. The unique characteristic of Spanish American mixing would be, simply, to produce these forms *more frequently*. In fact; even without the presence of abject poverty or noxious toxins, it is frightful just how many degenerates are born to Spanish American hybrids that are relatively inept for the propagation of the species; born to crossbreeds who, of Jehovah's initial commandment—"be fruitful and multiply"—only satisfy the first part . . . Amongst *the men*, we have the effeminate mulatto musician, a pianist with large knuckles, a virginal figure, a flute-like voice, and a romantic's coquettishness; and the Indian half-breed politician, with his beardless face and gelatinous eunuch's belly . . . Amongst *the women*, we have the spinster mulatta, so pleasant when she is an intellectual and makes for an excellent schoolteacher—which produces the misleading sensation that her race, if it would stand firm, would normally be apt for feminism—and then so repulsive when she is not an intellectual, but a simple seamstress who fills her empty heart up with any sort of fanaticism she can find, be it the icons of the temple or the speaking ghosts of the three-footed table . . .

I would call the phenomenon *degenerative semi-sterility of the human hybrid.* And do not oppose this idea by arguing that there has been mixing in all places in all time periods, mixing that, far from bringing about such a phenomenon, has even benefited certain races. Why must it arise in Spanish America, and not in Europe, Asia, Africa? . . . To this objection, two responses. First, the history of the ancient migrations of races is obscure, and as such it would not be difficult to comprehend that this phenomenon of relative sterility, were it to have taken place, has still not been demonstrated. Second, the confluence of races was more forced in America, less spontaneous than the rest of the instances in the historical record. It is hard to find another case of an immense continent populated by an aboriginal people and then conquered by foreigners from a completely different line, who, in addition, bring in a third and different race to work the land, followed by the crossbreeding of all of them together . . . Such an occurrence is artificial, anomalous, heterogeneous, and the consequent crosses are heterogeneous as well, as had never been seen before, as well as anomalous and artificial. Furthermore, they are brought about all of a sudden, due to complicated causes, such as the lack of white women at the outset, the concupiscence of certain tropical climates and Europeans' incongruous monogamy, the wealth, the poverty . . . It is brought about all of a sudden: I mean to say through external causes, before the black and white immigrants could adapt to their new climates. Quite differently, in the rest of the historical racial crosses, which I presume to be not only neutral, but even beneficial, the circumstances lead one to presuppose a prior adaptation to the geographic environment; these are *natural* cases. The case of the Spanish American races can be said to have been *artificial*, and even contrary to nature; and mother nature—who, as a woman, is vengeful—can be said to have exacted her revenge on them. How? By raising the average level of degeneration whenever they attempt to maintain the outlandish human hybridizations produced in defiance of her laws.

On the other hand, when crossbreeds tend markedly toward one specific physical type, be it White or Colored, they are extremely fertile. Anthropologists would say this is caused by "retrogression." The layman would find a different cause, especially in the case of people of European origin in whose veins there runs only a sixteenth or a thirty-second of African blood; they would say . . . by selection! Or, rather, by choice, by "chosen affinity." For it is the case that these distantly crossbred types have certain unique sexual charms which the *pur sang* lacks. Especially the females! What could be more graceful, what creature could be more exquisite than a European woman with faint mulatta traits, over whom civilization has passed, voluptuously, throughout several generations, its hand of lime all across her body, a female that brings together the beauty of a white woman with a certain spicy and attractive *je ne sais quoi* which is her own, like some exotic *nonchalance* of the harem? The force of her deep dark eyes, that legendary force, has been ardently sung by the poets of America. And there is a typical and elegant song that narrates, with adorable candor, the case of a dark-skinned or mulatta named Trinidad, a song which is very popular throughout the New World—I was about to say from pole to pole; but, to avoid such an overstatement, I'll say from Texas to Patagonia; and it is true! Only truly great precision and clear realism could have thus spread, over both sides of the Antilles, such a far-flung ballad as this one . . . Because the mulatta Trinidad is from Havana; there, according to the text, she was strolling through the streets one morning . . .

Two men came and grabbed her,
and took her to the jailhouse,
by an order of the law . . .

She must not have been much of a crook! . . . She cried before the judge, with the nervous tears of a princess taken prisoner . . . The judge, upon hearing her, and while seeing himself reflected in her eyes, "with no way around it" had such a ruthless *coup de foudre*, that, right then and there, he took her by the arm, walked her to the church, and . . . married her. This is what the popular song says. Where in Europe today do they sing of such a startling force attributed to a pair of female eyes? "My Indian girl" and especially "my Black girl!" are supremely sensual expressions in Spanish America.

The fact is that African blood joins together marvelously with Spanish blood, at least with that of Spaniards from the south, because, as I have said, according to modern anthropologists, the primitive Iberians were in part migrants from Africa, they were Afro-Europeans. This results in certain affinities that add character to the mixture. Not true in the case of Hispanics and Indians, as the latter are probably descended from a stock that has been isolated since the Quaternary. As such, one could call a Hispanic-Black cross simple crossbreeding; and Hispanic-Indian or Afro-Indian, true hybridization.

And when all three elements come into play—Indigenous, Spanish, and Black—the level of hybridization is such that degeneration will almost always be its immediate consequence, until the dissolution of the *false species* . . . One can then admit that Spanish American crossbreeding is not always viable; which gives us its relative sterility, which even Doctor Pangloss, the ultra-optimist philosopher, would accept, surrendering in the face of the evidence. Candide, his disciple, would ask him what good this anthropological phenomenon does for humanity . . . And frankly I cannot manage to discern just what justification could be provided by that good-natured, that terribly good-natured Pangloss, for his *tout est pour le mieux*, especially if it fell on him—just as all calamities fell on him—to play the role of the mule . . . Truly, reader, can you not understand the pessimism and even the perversity of the mules? And if "to understand is to forgive," forgive him when he kicks!

Moral sense in modern societies is largely the product of Christianity. We have a Christian morality. We inherit it, in the form of "mandates of conscience," as theologians say; of "innate ideas," as Descartes argued; of "categorical imperatives," as Kant would formulate; of "psychological predispositions," as positivist psychologists have termed it . . . However it is called, the fact is that all normal men, after so many centuries of Christian heritage, are born with a hereditary ability to distinguish between good and evil. Only degeneration can erase it, giving way to atavistic and antisocial types. This, in the case of Europeans. Not so, naturally, in those who do not possess this Christian heritage—or Buddhist, which could stand in for it—as is the case with American Indians and imported Blacks. As such, a distinctive and essential trait common to both Indians and Blacks, to mulattoes and Indian half-breeds, is the *lack of moral sense*, which is why among the Christian European peoples it is called "common sense."

Where the absence of the American hybrids' moral sense can be seen to be most harmful for society is in their lack of probity. This is why the characteristic phenomena of the Spanish American political economy are "rulers of blood and plunder," speculation on fanciful and improbable currency, individual and social bankruptcies, abusive and absurd taxes, unprosperous public companies . . .

In terms of sexual morality, African blood and even Indigenous blood, as natives of the tropics, of the polygamist regions, is much more "frequent" than European blood, which comes from cold climates and tends towards monogamy. The Colored populace in America is thus, due to ethnoclimateric reasons, necessarily less chaste, less "virtuous" than the Whites. And so much so, that in Rome on one or another occasion it was discussed whether or not an exception should be made that would allow the Spanish American Catholic clergy to marry, as, at least in the hotter countries, its members tend to carry abundant and susceptible African or Indigenous blood in their celibate veins . . .

[. . .]

OUTLINE OF RACIAL STRUGGLE IN THE ARGENTINE REPUBLIC

Imbued with the democratic tradition of the French Revolution and the constitutionalism of the United States, Argentinean historians have falsified Argentine history. In making up the idea of the great multitudes and principled political parties, they created a curious national mythology which, in the current treatises, passes for absolute truth.

They would have come closer to the truth had they left aside the French lyricism and the Yankee placards and applied to the events at hand any scientific viewpoint at all: *economics*, for example. For though the viewpoint of economics is not always sufficient to explain the facts of history, the great religious wars and legendary political passions, in Argentine history it is enough, because it lacks all such wars and passions. A new people which has had limitless and rich territories at its disposal, Argentina stands before Melpomene with her hands as white and her heart as pure as an adolescent virgin climbing the steps of an ancient pagan altar to be sacrificed . . .

The great internal movements in Argentine history can be reduced to three: the War of Independence (1810–1816); the provincial conflicts with the *caudillos*[6] (1816–1825); the interprovincial conflicts subsumed under the name "National Organization" (1825–1880).

1st. The War of Independence did not originate in lofty democratic ideals, nor was it carried out by multitudes yearning for glory and liberty. It was merely a movement initiated, without a thought to its future ramifications, by the bourgeoisie or criollo commerce of Buenos Aires against the irritating system of the Spanish monopoly. Its starting point was an energetic commercial protest, first formulated in the *Landowners' Petition* and later in the debates held in the City Hall from the 22nd to the 25th of May, a protest which, amply and unambiguously recognizing the sovereignty of Ferdinand VII, only attacked Spanish domination in so far as its economic excesses were concerned. The "people," out of ignorance and indifference, did not follow the "landowners" of their own volition; but the latter, as masters, would know just how to drag them along, if it were to become necessary, into clear and open rebellion . . .

The historical circumstances, the defeated British invasions, the alternating weak and willful politics of the metropolis, the romantic *venticello* of the French Revolution, the United States' example, all contributed to widen the torrent and channel it toward the democratic tendency praised by the eighteenth century's philosophical trends. And so it was that the primitive protest of the criollo bourgeoisie came to grow and take in foreign ideas, until proclaiming itself to be a "democratic revolution." This was an odd piece of forgery because, while there was a capable governing class in the colonies, there was no trace of a European and Republican people. It was a democracy without a *Demos*!

2nd. After Independence had been achieved, the aristocratic organization that Spain had given to its colonies persisted. The land, divided into latifundia, belonged to rich idlers of the cities, and the Colored populace of the countryside lived in the most miserable of conditions. This populace then initiated the second historical movement: it appointed *caudillos* and moved the political war into the cities . . . Such was the internal disorder produced, the confusion and the anarchy of the governments, that its high point has been called the *Chaos of 1820*. And just as the criollo bourgeoisie had triumphed in the War of Independence, in the provincial *caudillo* conflicts it was the peasantry that won.

3rd. With this victory the provincial governments were consolidated; each province had its *caudillo* of rural origin, that is to say, a *caudillo* whose all-embracing authority was based on his prestige among the largely Indian rural masses. But, unfortunately, wealth turned out to be distributed with utter inequality among the provinces: there was one very rich one, Buenos Aires, which counted its livestock in thousands of head of cattle, and many poor ones, very poor ones . . . So the latter, which would not consent to the domination of the rich province, waged constant wars against the province-capital. Against the men of the port's unitary ideal they raised the federalist standard . . . And the struggle did not end until a federal constitution was instituted and the city-capital was federalized.

And yet, though it may not appear so, these three great movements (the city-capital against the metropolis, the countryside against the cities, the poor provinces against the province-capital) hide and perpetuate a racial struggle which is latent and deep-seated, almost shameful. And it must be recognized that, in this struggle, the richest ethno-geographical factions were almost always the ones with the least amount of crossbreeding. I will describe this struggle here because I hold it to be typical of so many of the Spanish American republics' political and economic conflicts.

1st. In the first movement, the crossbred *criollo* fought against the Spaniard, who he quite graphically called *Goth*, which is a blond and racially White man. It is a fact that Juan de Garay founded the city of Santa María de los Buenos Aires in 1580 with eighty-eight soldiers, of which only nine were Spaniards and the rest were "criollos," the sons of Paraguayan Indian women.[7] According to approximate calculations made by Sarmiento, people with purely Spanish heritage made up barely a sixth of the city's population in 1776, given that following the importation of African slaves most people had at least a drop of African blood in their veins. Around the end of the eighteenth century the portion of the population that *called itself* "Colored" constituted around *half* of the census,[8] a group made up almost entirely of Blacks, and mulattoes who were dark enough that they registered as Blacks, since there were very few full Indians who could withstand life in the city.

The ethnic levels put forth by the Buenos Aires census, far from being exceptional or surprising, are just one of so many such cases in the founding of colonial cities: the criollo was the plasma of nearly all of them, but in the most important ones there arose a Hispanicized governing class made constant by immigration; a rich and proud bourgeoisie, which in the cities of the Viceroyship of the Río de la Plata called itself *the decent people*, as opposed to the *rabble*, the Colored mob that they shunned.[9]

2nd. The second movement in the internal history of the Argentine Republic, that of the outlying and rural masses against the rich urban classes, can also be interpreted as a new racial conflict: the Indian half-breeds of the countryside against the Europeanized criollos of the cities. The latter, forgetting their crossbred line of descent, became the successors of the "Goths" in their racial pride, and, just like the "Goths," they were dealt the revenge of still more criollos who were still more numerous.

3rd. Not even here, upon the victory of the semi-Indian rural gauchos, did the racial conflict end. In the third movement, that of the interprovincial struggles, the poor provinces of the interior represented a more or less Indian ethnic tendency against the Spanish ethnic tendency of the province-capital. Given its cooler climate, its geographic positioning around a port, and the lowliness of the tribes on the pampas, the city of Buenos Aires and its surrounding plains, which received a constant flow of European immigration, experienced a relative and growing distancing from the Indians during the colonial period. Furthermore, alcoholism, smallpox, and tuberculosis—God bless them!—had decimated the indigenous and African populations of the province-capital, purifying its ethnic elements, Europeanizing them, Hispanicizing them. On the other hand, in the interior, the tropical temperatures, the difficulties of communication with the metropolis, and the culture and beauty of the indigenous races spread and increased crossbreeding up to the period of independence. This is why there is also an ethnic factor to the new interprovincial struggles that took place just after *caudillo* system had been consolidated.

[. . .]

NOTES

1. TN: The word that throughout this chapter I have translated as "crossbreeding" is *mestizaje* in the original Spanish, and "crossbreed," *mestizo*. The word *mestizo* has of course found its way into the English language to describe people of mixed Spanish and Amerindian descent. However, given that here it is used in the broader context of the crossbreeding of biological varieties, I have opted not to use it.

2. TN: *On the Origin of Species*. Chap. 1, Sec. 2, passim (appears in Spanish translation in the original).

3. TN: Ibid.

4. TN: See introduction, note (iii) for an explanation of the use of the world "criollo" in South America.

5. TN: In Argentina "Chinese" (*chino* in Spanish) is often used as a derogatory term for indigenous people.

6. TN: See introduction, note (xvi) for an explanation of the term *caudillo*.

7. OAN: J. L. Cantilo, *Juan de Garay*, Buenos Aires, 1904, p. 176.

8. OAN: D. F. Sarmiento, *Conflictos y armonías de las razas en América* ("Racial Conflict and Harmony in America"). Vol. II, articles, "Introduction" and "The population in the capital of the Viceroyship."

9. OAN: Nevertheless, it is worth noting that the population of Buenos Aires was ethnically one of the most Europeanized, if not the most Europeanized in all of Spanish America. The census conducted in 1778 by the viceroy Vértiz shows that, out of a total population of 24,205 inhabitants, Buenos Aires had 15,719 whites

(Spaniards and criollos), 7,268 blacks and mulattoes, and, lastly, 1,218 Indians and Indian half-breeds. Later statistics show even greater proportions of whites . . . But we must not forget that even though the most European of Spanish American populations, that of Buenos Aires, is clearly made up mostly of criollos, these criollos that were registered as whites in the census were the descendants of Spaniards and Indians. The label "whites" has fooled many would-be sociologists. In reality they were "whitened," by European crossbreeding and city life . . . In any case, this fraction of indigenous blood has been diluted so much that in the twentieth century it really can be considered trivial. This is the result of the abundant immigration from Europe that mixed with the population starting in the final decades of the nineteenth century.

THE SOCIOLOGICAL EVOLUTION OF ARGENTINA

From Barbarism to Imperialism

(1909)

José Ingenieros

INTRODUCTION

Humankind is a biological species that evolves according to laws that sociology attempts to understand. The Argentine nation is a group of individuals of this species; under determined circumstances of time, mode, and place these individuals constitute a social aggregate which fights for its life—under ecological conditions common to all of them—with other groups of the human species and with the rest of living creatures.

How might we trace the course of Argentina's sociological evolution? Or, in other terms, how might we order into a systematic series those principal variations of the Argentine social aggregate from its formation up to our times, elucidating its causes? A general—and therefore philosophical—point of view is foreign to the analytical and descriptive concerns of chroniclers and historians, regardless of whether one makes use of their work in cases where they yield true facts and not when they translate their subjective feelings or beliefs, which is commonplace. This aside, as the reader will have gathered from our preceding essays, sociology can only consider a social aggregate to be a part of the human species that lives and reproduces on the surface of this planet that we inhabit.

As it is a living species it is subject to biological laws; as it is capable of living in social aggregates it is subject to sociological laws that depend on the former; since it is able to transform and utilize the natural energies that exist in the environment in which it lives, it evolves according to economic laws, which are specialized laws within the preceding two.

All knowledge should be considered contingent and provisional: truth—understood to be the agreement between ways of thought and the data of experience at any given moment in universal evolution—is constantly coming into being, since the phenomena that reality presents before our senses are innumerable. One can only demand of doctrines that they not be in contradiction with the experimental data of their time period, and their importance can be measured by the quantity of homogeneous facts that they allow one to systematize or explain. Men form part of a world in which reality evolves alongside interpretations of it.

In studying Argentina's sociological evolution from this point of view, it is evident that the opinions expressed further along cannot correspond with the trends of any political party or of any such chronicler. There is nothing that would lead one to believe that such a circumstance might greaten or diminish the authority of what is written here. The interpretation of sociological experience has never been the norm in collective political action; chroniclers have reflected their own personal sentiments or those of their immediate group, subordinating the facts to their political or aesthetic interest, if not the natural inclinations of their imaginative temperament.

Sociological changes tend to take place without communities' realizing the direction of their own itinerary. The species, races, nationalities, parties, groups, individuals, are all pulled along by natural necessities that engender sentiments and are reflected in useful beliefs. The sociological notions that manage to come into being independently of these beliefs may sometimes come close to an objective understanding of phenomena, but they do not determine them; scientific understanding is born of experience, much like the surface of a still lake reflects the image of a reality that exists independently of it.

Just as consciousness takes up only a minimal portion of an individual's total psychological activity, in the realm of sociology communities tend to act with a limited understanding of the ulterior contingencies of their actions. The struggle for life between human groups is carried out unconsciously in the majority of cases, without knowledge of the orientation of this collective effort moved by useful beliefs. Still, the fact that it is unconscious does not make it any less intense. Social groups tend to be like vessels sailing without a compass, pulled along by currents whose secret resides in geological and biological causes which social consciousness never suspects.

This is why some of the conclusions put forth in this synthetic construction will surely contrast with many ideas accepted through habit or mental inertia; they stray from norms consecrated by routine, always stubborn in the face of any new synthesis, even if the latter is founded on prior analysis of a reality which is constantly coming into being.

In order to deduce from facts the most general laws governing them, and the most constant relationships that hold between them, scientific thought makes use of two intellectual activities whose results are complementary. Sociology can take no interest in understanding particular facts unless it is to determine the general laws of which they are an expression; the sociologist is not a collector of data, but rather their interpreter. If one has not been able to use abstraction and generalization to differentiate between the facts and then group them according to their similarities, a chronicle constitutes mere empirical experience and not scientific understanding. The analysis of facts and the synthesis of this analysis are the two intellectual processes that are necessary in order to systematize the data provided by experience. The analytical inclinations scrutinize and prepare materials that will be brought together later on by the synthesizing inclinations. Objective and meticulous examination of partial phenomena constitutes the first step in understanding, historical narration. Fusing them into synthetic generalizations represents the ultimate goal of the process, sociological interpretation. An analysis which is not completed by synthesis is an incomplete function, while synthesis without sufficient prior analysis tends to be both precarious and sterile. The result of the crossflow between the two operations is a balance that brings the constructs of thought closer to the data found in reality. In order to formulate this synthetic conception of Argentina's sociological evolution, the author has taken into account the analytical data set down by historians, and has criticized the sociological essays written in this country on determined periods, events, or personages.

Thinking outside the constraints of race, nationality, class, or party—which is easier in an unstable and heterogeneous environment such as Argentina—the present essay attempts to show the apparent antinomies that develop around the following two directions: the evolution of indigenous barbarism towards the economic regime of capitalism (on the internal plane) and the evolution of the colonial establishment towards imperialism (on the international plane). Far from an improvised or premature generalization, it is the systematic expression of ideas that have been slowly constructed and fragmentarily exposed in critical studies, of very unequal merit, over the course of ten years [. . .].

The three chapters into which this sociological synthesis are divided revolve around new fundamental concepts. The first establishes a general perspective that resolves the apparent conflict between the two most firmly established sociological theories; in the following chapters this perspective is applied in the examination of the struggle for life between the groups that agitate within the social aggregate, and between this and all other human aggregates.

1st Resemblance between economic sociology and biological sociology, till now considered antithetical systems.

2nd Economic interpretation of Argentina's political evolution, from the initial manifestations of colonial life to the present formation of nationhood.

3rd Explanation of imperialist politics according to economic sociology, regardless of the common divorce between the supporters of one system or the other.

The first concept brings one to view national and international politics as simple manifestations of the struggle for life between social aggregates.

The second allows one to get an idea both of the evolution of internal politics in harmony with the current interest of the capitalist economic regime that is unfolding in racially White nations, as well as the evolution towards the progressive socialization of great collective functions into the hands of the state.

The third allows one to make out the future tutelary actions of Argentina over the countries of South America, and the likely orientation of its international politics towards a pacifistic imperialism.

PREMISES FOR AN ARGENTINE SOCIOLOGY

1. The formation of Sociology. 2. Sociology as a natural science. 3. The sociological workings of national and international politics. 4. The two problems in the Sociology of Argentina.

1. *The formation of Sociology.* The evolution of scientific thought has not been uniform. The progress of the historical disciplines—which tend to be easily confused with sociology once they turn scientific—have not run side by side with the development of the physical and biological sciences. The reason is obvious: in universal evolution social phenomena rank below cosmic, geological, or biological phenomena. The study of man in terms of his most highly evolved phenomena, which is to say his individual and social psychology, necessarily comes after the study of the physical, chemical, and biological phenomena that preceded his genesis and transformations.

   History, as with all branches of human understanding, comes about progressively. Bernheim has summed up the question by pointing out three distinct principal phases. The first, narrative or expository, merely attempts to lay out the facts at hand. The second, instructive or pragmatic, coordinates the narration of the facts so that they converge upon the demonstration of a determined thesis; this is often unilateral. The third, evolutionary or genetic, attempts to explain what has determined the historical phenomenon, its meaning, and it relationships to other prior, concomitant, or consequent phenomena. The first only takes up information and relationships extrinsic to the facts; the second, those intrinsic to them; the third, both of them equally.

The idea of history has been faulty for centuries. From mythological interpretations characteristic of all primitive peoples, humankind fell into theological and mystical systems, which are equally absurd; Bossuet was their most genuine representative. Later on individualist theories of history came into bloom, attempting to show that history was simply the result of the intelligence and will of a few brilliant men: this point of view was carried to its extreme by Carlyle, Emerson, and Mackaulay, and brought about another error: biographical history. Against this there arose historians and philosophers of merit, who saw the attribution of too much influence to heroes and representative men to be erroneous, since these men are nothing more than the natural products of the environments in which they appear, the condensers of the needs and aspirations that are present in the entire people; Buckle and Taine laid firm foundations for this school. But it was soon observed that all of these theories were exclusive or apriorist. History could not get bogged down in any of these conceptions, nor was it under any obligation to remain aloof from the flow of contemporary science down the wide conduit of deterministic evolutionism. After being brilliantly applied by Laplace to cosmic phenomena, by Lyell to geological phenomena, by Lamarck and Darwin to biological phenomena, Comte and Spencer tested it out on social phenomena. Already in the intuitions of Schelling, Hildebrand, Guizot, Tierry, Quetelet, Thomson, Buckle, Taine, and others, it had started to become clear that man was, foremost, a living animal with material needs that he had to satisfy by subsisting off of the environment in which he lived. This perspective highlighted the economic basis of historical evolution, which was formulated in some of Marx and Engels's essays, and acquired the workings of a system in the writings of Aquiles Loria.

It is impossible not to admit that there has been a movement toward an interpretive synthesis of human evolution, which was once the realm of historical philosophy and now belongs to sociology. An undeniable progress separates us from the initial theological interpretations and from all of the purely metaphysical systems that followed; among them we can include the idealist or intellectualist conceptions, in the vein of Hegel or Comte, and the theories that have been called materialist, such as those spread by Buchner, Moleschott, or Vogt. All of these were conceived outside of the scientific method, and all were metaphysical; they went from intelligence to phenomenon and not from phenomenon to intelligence. In this sense positivist philosophy and the nascent field of sociology were unable to rid themselves of the philosophical methods and trends that preceded them.

But scientific history, which is to say sociology, did not stop there. The study of human evolution got under way with more or less dependable methods, though from partial points of view. Each school, each author, sees just one facet of its complex prism and is inclined to see all the rest as subordinate. Thus Buckle, unable to part with a certain measure of intellectualism, subordinates historical evolution to the influences of the physical environment. Others, such as Kidd, and in part Le Bon, consider the phenomenon of religion and its transformations to be fundamental. Demolins ascribes supreme influence to topography, creating geographical sociology and seeing in great social roads the cause behind sociological prototypes. Ardigó understands the essence of human history to be the evolution of the phenomenon of law. Others, such as Lapouge or Gumplowicz, subordinate everything to race and the struggle for life, or to anthropological factors in a range of senses, as with Simmel or Folkmar, etc. In short, two vast schools vie for eminence in modern sociology. On the one hand there are the organicists, among them Worms, Lilienfeld, Schaffle, and Novicow, determined to view human societies as organisms and endeavoring to apply to them the laws of biology. On the other hand there are the economists, such as Loria and De Molinari, who attempt to reduce sociology to problems of political economy.

Are they all wrong? Probably, although some of them more than others. But each one of them has provided important materials for the overall task; a grain of sand here, a solid block of granite or a powerful column there. This is why the science of history, while still not on par with chemistry or cosmography, is much more than some mere alchemy or astrology.

The evolution that has been brought about in how history is studied makes it impossible to disregard the fragmentary importance of contemporary sociologists' vast body of work, even if it may be considered one-sided and incomplete. The sociological disciplines, though they give us no definitive guidelines for studying the evolution of Argentina, do offer us some fundamental conclusions and solid standards of interpretation which, when put into practice, will allow us to extract from the habitual narrations of history certain general principles, which will be less and less imprecise.

2. *Sociology as a natural science.* Prior to outlining a genetical interpretation of the evolution of Argentina (particular phenomenon), it is important to set out a few perspectives that can be applied to the whole of human evolution (general phenomenon). To do this it will be necessary to forgo all apriorism or telic preconceptions, all prejudices in favor of one principle or of one theory.

Support for the study should be gathered with the methods common to all the natural sciences. When studied objectively, sociological phenomena are only the combination of a series of physical, biological, and psychological facts, which is the same as what the rest of the sciences study. In this sense one could specify the following formulas, which are the synthesis of a complex conceptual framework.

Man is not an aerolite fallen to earth through some whim of the supernatural forces; he is a complicated manifestation of life, which is in turn a manifestation of material and of universal energy. Man is a living creature, nothing more; in him life assumes an infinite number of intricate manifestations, but without escaping the common laws of biology and its general principles. Just as with the rest of living creatures, he fights for his life in order to satisfy basic and indispensable needs: the preservation of the individual and the reproduction of the species. Humanity, viewed as a biological group, has no mission whatsoever to carry out in the universe, just as fish or weeds have no such mission. This lack of final cause excludes the existence of absolute moral principles. The spring that sets man's social activity—his conduct—in motion is the sum of his needs; his consciousness of the latter—subjected to rigorous determinism—is merely the apparent motive behind all individual or collective actions.

This initial point of departure is in line with the ideas that are common to all supporters of historical economism, understood in its broadest sense: the material necessities of life are the motive behind the activities of the human assemblages studied by sociology, forgoing all transcendental ends.

Yet, at the same time, it shows us that the basis of economic factors lies in the purely biological needs of man, considered as a species. Humanity's basic needs are the same as those of any animal species, and those of every aggregate or sociological group are equivalent to those of any grouping of living creatures, be it a colony of microbes, a hive of bees, a herd of horses or a tribe of men.

In reality, the very conditions of human progress have developed certain elements that are essential in the struggle for life, among which the greatest is the need to produce its own means of subsistence. And yet, this development—which can be considered characteristic of the animal species to which we belong—is only a higher, more evolved form of the tendency to satisfy fundamental needs, which is common to all living creatures. The essence of economic factors is constituted by the purely biological needs of man, seen as just another branch of a polymorphous phylogenetic evolution.

These considerations—which, though synthetic, are no less decisive and clear—push us to pose a second premise which resolves, in a novel manner, the doctrinary conflict that has unsettled the whole of contemporary sociology: "organicism" and "historical economism," far from being opposing conceptions, are two partial ways of understanding the genetical problem of sociological formations. There is room for both of them in a biological theory that views social aggregates as mere groups of a biological species, humanity, and not as "superorganisms."

Within this sociological perspective, sociology is a mere natural history of the groups into which humanity is divided; it is the most complicated chapter of zoology and that which is of the most interest to us. The dissonance between organicism and economicism can be reduced to simple terms.

The biological needs shared by all living species, the human species included, give rise to a range of phenomena governed by laws that are presided over by evolution and the struggle for life, laws understood in the broadest sense: and this is the point of view upon which biological sociology is founded, a natural science that should replace Spencerian organicism. But these very fundamental needs are modified in the human species by a heightened level of mutual affiliation in the struggle for life, adding certain traits that constitute the development of economic phenomena along with the creation of new relationships between the groups that make up the species, between the classes that make up the groups, and between the individuals that make up the classes. This is the perspective on which sociological economism or historical materialism are founded.

We biological sociologists elevate the problem to its general dimension, which is biological; economistic sociologists face it in its particular dimension, which is human. But the essential phenomenon that presides over the whole of social evolution is one: the needs that human aggregates must satisfy in order to meet their double biological aim, the preservation of the individual and the reproduction of the species. Political economy is the application to the human species of the basic biological laws governing the struggle for life.

In this sense—unanticipated by sociologists of either school—we believe that historical economism can be interpreted as the simple application of biological sociology to the evolutionary study of human societies. Let us look for some corroboration of this doctrinal and general point of view in our own social context.

We may, then, repeat that biological sociology, understood to be a natural history of the human species (in contrast with primitive Spencerian "organicism"), and economistic sociology (historical materialism), far from being opposing conceptions, are partial but concordant ways of understanding the problem of social aggregates' genesis and evolution.

3. *The sociological workings of national and international politics.* Since man is a living creature who must satisfy material needs in order to preserve and reproduce his own life, just as with all other biological species, the existence of social groups is subject to similar contingencies of primal importance.

The historical path of every people passes through particular evolutionary phases due to the intrinsic conditions of the double natural-economic environment in which it develops. This mold produces the diverse organizational forms taken on by each human society and is the substratum that nourishes and holds the roots of the political, legal, moral and other such institutions that constitute the superstructure of these societies at any given moment of their evolution.

This naturalistic interpretation of history, which in the realm of scientific sociology we prefer to the different theological and idealistic interpretations, leads one to view it as a conglomeration of phenomena linked together by inevitable causal relationships and not by ends independent of both the world and of life. Each historical phenomenon has determining factors which could not have done otherwise than to produce it, and it, in turn, inevitably determines other historical phenomena. This perspective, at which contemporary science has arrived with its most recent interpretations, makes clear that social facts must be studied in terms of their objective nature and as truly complicated manifestations of the living matter that evolves on the surface of the planet we inhabit. Any attempt to investigate their metaphysical transcendencies is as vain as allowing for the participation of fate, of chance, or of an unknowable end that might have presided over the creation of matter and of life. For science these factors have no positive importance.

Humanity is simply the case of an animal species fighting for its life in a limited physical environment, namely the earth. This species has varieties, which are the races, and the environment in which they fight for their lives is heterogeneous. These fundamental causes determine geographical and ethnic differences that are primary factors in the constitution of all social groups. Due to the interference of other innumerable factors these groups evolve and form nationalities, which are transitory social aggregates in the face of infinite time.

If one attempts to take in, with one broad gaze, the full panorama of the diverse activities carried out by men who live in societies, it is readily apparent, even for the most superficial of observers, that the Darwinist principle of the struggle for life continues to govern the social world, though it undergoes the important modifications that we have pointed out. As a biological species, man is subject to the rigors of a struggle for life against the plant kingdom, and against the rest of the animal species. As an animal prone to affiliation with aggregates or groups, he is subject to new types of struggle: be it as a member of a social aggregate, be it as an individual.

There are three possible types of struggle between men: 1. Between social aggregates; 2. Between aggregates and individuals; 3. Between isolated individuals. Two nations that destroy one another in a war for economic supremacy find themselves in the first scenario. A delinquent who commits antisocial acts is an instance of the second. Two savages who fight over an edible root find themselves in the third.

The different forms that the struggle for life can take between social aggregates—as well as between the collective groups that live within each aggregate—vary *ad infinitum*. Their reciprocal relationships are constantly changing, given the persistent goad of their conflicting interests. A primary cause behind conflict arises from ethnic inequality, as in the racial struggles studied by Gumplowicz, Ammond, Lapouge, and Winiarsky. As history evolves they diminish, because these kinds of conflicts subside as the races tend to come together under the hegemony of those that are best adapted to the struggle for life, as Colajanni and others demonstrated some time ago. Within the same race the diverse conditions for economic development due to the influence of the natural environment determines the formation of diverse political aggregates. Different states are formed, and there arise between them conflicts and interests which are the cause of fighting between nations; one need only recall Novicow's studies. The differences between the social functions of each sex, and the superior needs of preserving the species, determine the struggle between the sexes, brilliantly analyzed by Viazzi, as each one endeavors to exercise a greater authority over the other, and capture the right to love at the cost of the least possible effort. Within each social aggregate the division of labor determines the appearance of social classes that, over time, can come to have opposing or divergent interests: this is how class struggles appear in the studies of Marxist sociologists. From another, narrower point of view, the common interests of those who live off of the practice of a common skill or function bring about a struggle between them and the rest of society, in forms that vary from the moral complicities of an *esprit de corps,* to capitalists' and proletarians' economic syndicates. One could point out one hundred specific forms of the struggle for life amongst communities, since whenever there exists a common interest, be it permanent or transitory, there is a collective struggle against the rest of the social aggregate or one of its parts.

In those social aggregates that are already evolved and organized into nations, all of these forms of the struggle for life are polarized around two principal manifestations: 1. Conflicts of interest between the different groups that compose a single social aggregate (internal politics); 2. Conflicts of interest between the different social aggregates that coexist in time and are delimited in space (international politics).

As such, this problem can be posed in the following terms:

National politics is the struggle for life's expression among diverse groups with heterogeneous interests, but subsumed under the common interest that serves as the basis of nationality. International politics is the struggle for life's expression among different social aggregates that have evolved to the point of constituting separate nationalities, with the participation of the physical environment and of race, or without it.

4. *The two problems in the Sociology of Argentina.* In addition to its particular advantages for the social aggregate we are studying, the genetical study of the most recently formed social aggregates would be of benefit to sociology generally. Their evolutionary development reduces, to a short space of time, the transformations that have lasted centuries in other social groups. It is one particular case of the general principle established by Aquiles Loria, according to which in the economic evolution of the contemporary colonies we can observe a condensed version of the evolution of States, whose civilization dates back to ancient times and has come about slowly. Moreover, this sociological principle is nothing but an application in the sociological world of a law demonstrated by Haeckel in the biological world: the ontogenetic evolution of the individual reproduces the entire phylogenetic evolution previously undergone by the species.

There is no reason for the history of America to admit a priori, in regards to the origin of its pre-Columbian social groups, the monogenist hypothesis of Spencer—which, though accepting the multiplicity of its primitive societies, attributes to them one common character of relative homogeneity or imperfect differentiation—nor the polygenist hypothesis upheld by Tarde and Gumplowicz, which supposes an absolute diversity in social forms from the time of their origin. To these theoretical opinions we prefer the admirable ideas about the origins of man held by our eminent paleontologist Florentino Ameghino. Let it suffice us to know that, in general, in order to attain a state of civilization all peoples must pass, in their process of successive integrations and disintegrations—which Saint-Simon would call critical periods and organic periods—through the various stages of savagery and barbarism, the succession of which is determined by the different systems that characterize the means of production in each form of social organization (Marx, Engels, Loria). This economic factor, derived from fundamental biological needs through a higher level of evolution attained only by the human species, is that which principally determines the morphology of all social institutions, from the most rudimentary institutions of the marauding hoard to the most complex ones of industrial capitalism.

Nevertheless, though the stages of the evolutionary process are generally the same for all social groups, the phase in which each of them finds itself is entirely variable according to time and space, since the cosmic and artificial conditions of the environment are not simultaneously the same in all inhabitable parts of the earth's surface.

The Argentine social aggregate still finds itself in an unceasing period of formation; a century of independence has not been enough to definitively organize the institutional life of the Republic. The reason for this is simple: no social aggregate can attain a definitive organization for its political regime without beforehand having consolidated the economic structure that serves as its foundation, be it the appropriation and exploitation of free land, or the progressive development of industrial capitalism.

[. . .]

RACIAL CONFLICT AND HARMONY IN THE AMERICAS

The Indigenous Races and the Black Race

(1899)

Domingo F. Sarmiento

In order to write the histories of Peru and Mexico, the historian Prescott had at his disposal all the books, chronicles, and notes of those who lived at the time of the conquest in both of the two countries. Since these texts come from settlers located exclusively in Peru on the one hand, and settlers located exclusively in Mexico on the other, it is Prescott alone who reveals the perspective that a knowledge of both conquests has given him, which lends added weight to his opinion. Perhaps this knowledge is inflated by his own credulousness, so that he is not sufficiently critical of the information that the authors transmit to him as first-hand witnesses, be it in terms of the plausibility of the events they recount, be it in terms of the size of the armies in battle. Another more recent historian from the United States, Wilson, in his *New History of Mexico*, in demonstrating the material impossibility of a large number of the recounted events, says:

"I have presumed to doubt that water ever ran up hill, that navigable canals were ever fed by 'backwater,' that pyramids (*teocalli*) could rest on a foundation of soft earth, that a canal twelve feet broad by twelve feet deep, mostly below the water level, was ever dug by Indians with their rude implements, that gardens ever floated in mud, or that brigantines ever sailed in a salt marsh, or even that 100,000 men ever entered the mud-built city of Mexico by a narrow causeway in the morning, and after fighting all day returned by the same path to their camp, or that so large a besieging army as 150,000 men could be supported in a salt-marsh valley, surrounded by high mountains."[1]

Ondegardo, cited by Prescott, says the Indians' only tribute was human labor, since they had no other possessions.

"In [. . .] México and Perú [. . .][2] there was the same inability to diffuse even that scanty civilization which they really possessed; there was the same utter absence of any thing approaching to the democratic spirit; there was the same despotic power on the part of the upper classes, and the same contemptible subservience on the part of the lower."[3]

"In the most essential particular for which history can be studied, namely, the state of the people, Mexico and Peru are the counterpart of each other. For though there were many minor points of difference, both were agreed in this, that there were only two classes—the upper class being tyrants, and the lower class being slaves."[4]

"Under this extraordinary polity," Prescott continues, "a people advanced in many of the social refinements, well skilled in manufactures and agriculture, were unacquainted, as we have seen, with money. They had nothing that deserved to be called property. They could follow no craft, could engage in no labour, no amusement, but such as was specially provided by law. They could not change their residence or their dress without a license from the government. They could not even exercise the freedom which is conceded to the most abject in other countries, that of selecting their own wives."[5]

As far as the character and natural dispositions of the Indians in the period prior to the Independence of this Continent are concerned, don Juan de Ulloa, who traveled across much of America studying the situation of the colonies, makes the following assessment:

"The inclination toward leisure and indolence is the same amongst the Indians of Louisiana and Canada, as with those of Peru and the southerly parts of America, be they civilized or heathen; and the only practices with which the ones who live freely occupy themselves are hunting and fishing, which is the same as what we see among the nations surrounding Buenos Aires. In the pampas of the province with this name, it is the women who are in charge of sowing their few small plots of corn and a bit of squash (of the *zapallo* variety), who grind the corn in order to cook it according to their mores, and who prepare their traditional drinks, in addition to taking care of the children, because in this matter the fathers do not trouble themselves."

As if to back up this very assessment with more recent evidence from different parts of America, the French agent in Caracas, M. F. Depons—who published a diary of his travels to the eastern part of Terra Firma in South America, an account confined to the territory of the Captaincy of Caracas in the years 1800 to 1801—captures the characteristic traits of that country's subdued Indian populations in the following terms.

"The Indian is distinguished," he says, "in a most singular way, by an apathetic and indifferent nature which is not to be found among any other group. His heart does not beat even in the face of pleasure or hope; it is open to fear alone. Quite the contrary of human daring, his character is distinguished by the most abject timidity. There is no spring in his soul, nor liveliness in his spirit. As incapable of comprehension as he is of reason, he spends his life in a state of stupid unconsciousness that demonstrates his ignorance both of himself and of everything around him. His ambition and his desires never reach beyond his immediate necessities.[6] . . .

"All of a lawmaker's efforts to inspire in them (the Indians) the desire to improve their native faculties have miscarried. Neither the good treatment they have received in being admitted into society, nor the important privileges with which they have been favored, have been enough to wrest from them their fondness of that savage life, despite the fact that today they know this life only through their folklore. There are precious few civilized Indians who do not sigh over the solitude of the forests and who do not take advantage of the slightest opportunity to return to it.

"This is not because of any love of liberty, but rather because the find the shady abode of the forest more suited to their melancholy, their superstition, and their absolute disdain for the most sacred laws of nature.

"So accustomed were the Indians to lying, and so little did they feel the sacred obligation to tell the truth, that in order to prevent the misfortunes that their false testimonies visit upon the innocent, the Spanish have had to issue a law stating that no less than six Indians may be used as witnesses in a case, and that the testimony of these six beings is equivalent to the sworn testimony of just one White."

[. . .]

AMALGAM OF DIFFERENT-COLORED RACES

This was still one of the characteristic traits of Spanish colonization, which in this respect followed a different behavioral norm than that followed in the North by the Anglo-Saxon colonizers, about whom we shall speak elsewhere.

The Bishop of Chiapas's excessive philanthropy—which was roused by the cruel acts that the Spanish conquistadors carried out against Indians with the character and intellectual state that we have already described—misguided the Spanish into introducing Black slaves from Africa in order to replace the Indians in the forced labor of the mines and other American toils. Wilson, the historian of México, casts doubt on Father Las Casas's inflated statistics about the victims of his compatriots' cruelty toward the Indians. Meanwhile, since England took the lead in the modern crusade to free the Black race, like the Bishop of Chiapas before it with the Indians, English historians have raised to twenty million the number of Blacks transported between Africa and America. This figure does not include those thrown overboard every year by the thousands, after dying of nostalgia, hunger, and illnesses along the way.

"A forest life, a forest atmosphere, was their wont, together with such a living as that forest had furnished for generations untold. The beaver and the elk are not more certainly doomed by the destruction of our forests, than were and are that race of men, who were set apart in the primitive divisions of mankind as its lords. Like caged lions, a few generated and a handful survived the revolution in their mode of life—enough to evidence what once existed—enough to prove too that a race differing from our own in its organism were the primitive inhabitants of the whole of this continent. Spanish crime and Spanish cruelty has indeed destroyed thousands, but the wasting of the forests of the table-land its tens of thousands more."[7]

"This principle in the constitution of the human family was not understood by Las Casas. [. . .][8] His brethren, the monkish missionaries, in after ages reached the true cause and the remedy, empirically. The Indian, compelled to abandon forest life, was subjected to task-labor. If he survived the hardships of his new condition he became the progenitor of a family of agriculturalists (*Pueblanos*)—a race that is now repeopling Spanish America."[9]

One must keep in mind this important distinction between the indolent and rude aborigines, and their present-day degenerated descendants, who have suffered centuries of servitude.

Whatever was to become of it, the Black race entered as an alloying element into the metal of which the American people was to be formed. This came about when, after the destruction of the barriers that had divided the people into casts like in India and Egypt—and perhaps as in the Empires of México and Perú, as their historians would have it and as Bukle claims in his *History of Civilization*—they were called upon in virtue of their numbers to express their common will through the ballot, or some other manner. And yet there it is a fact which we will note in passing, whose consequences must still be dealt will centuries on: the Spanish nation—which had the bonfires of the Inquisition raging back in Spain so as to deprive its inhabitants of all contact with the new ideas that were brought along with the Renaissance, as well as the advances in the scope of geography and astronomy—did not, for three centuries, allow even a single foreigner or man of a race, ideology, or creed different from that of the Spaniards of those times to enter its vast American domains, after having subdued the Moors in Granada, the Italians in Naples, the Belgians in Antwerp, where all of the riches amassed in the Indies were looted in six days.

The world was about to see what would happen when one mixes pure-blooded Spaniards, the European element, with a heavy sprinkling of the Black race, diluted by an enormous mass of natives, prehistoric men, of low intelligence, and all three elements are almost completely unaccustomed to the practice of the political freedoms that constitute modern governance. However, these different-colored races did not form one homogeneous whole, as was the case with Gauls and Romans, Saxons and Normans, Germans and Lombards, Goths, etc., and even Arabs and Saracens, all of whom are all, in the end, varieties of one and the same race: the Caucasian. Agassiz denies that the offspring of Black and White, of Black and Indian, and of Indian and Black, by which we get "Mulattoes," "Half-Breeds,"[10] and "Mamelukes," can be produced without returning to one of their original prototypes. Still, common speech has gotten ahead of science in distinguishing between these diverse origins and their intermediate half-castes, which are still quite perceptible in Perú and Bolivia, though they are happily not so visible in our own Argentine society.

Mr. Brackenridge, secretary of the United States' mission from 1817 to 1818, who has left us the most extensive collection of notes on the happenings of the time and on the outward appearance of society, gives us a curious assessment of the numerical distribution of the races and half-races among the populace of the Upper Peruvian provinces.

"After deducting from this number, one-fifth for the monastic orders and the Old Spaniards, with their adherents among the nobility, there will be left between three and four hundred thousand ready and willing to support the cause of independence, exclusive of Indians, who may safely be counted on when they can consult their inclinations; but whose extreme ignorance and the slavish state in which they have been kept, render them of small account, compared to their numbers."[11]

In the way of a memorandum, which suits our purpose, he says elsewhere: "The great proportion of submissive and timid Indians in the population of these countries, tends much to favour the Spaniards, and gives them advantages over the patriots, who, through policy, if for no other reason, adopt a different cause. The Indians are continually impressed into the Spanish armies, and accustomed, as they have been for centuries, to the most abject obedience and slavery, they not only tamely submit to their fate, but even make excellent soldiers."[12]

Returning to the races' proportions, Brackenridge adds: "From the foregoing estimate, it will appear, that the proportion of whites to the aborigines, is about one to five. But even amongst those regarded as whites, or Spaniards, the proportion of mixt race must be very considerable, a circumstance which tends to efface the line of distinction between creoles and the natives, and to bring them nearer to each other in point of interest and feeling."[13]

"The next class in point of numbers, is that of the Mestizos and Cholos. The first, spring from the mixture of the white and Indian, but not so far removed from the Indians, as to be ranked in the class of the Spaniards, although in their dress, manners, and language, no very essential difference is perceived. They can generally read and write; they carry on the small traffic of the country, are clerks and agents of one kind or other; they manage the large estates of the wealthy, but are seldom possessed of more than a moderate share of wealth; they are, of course, in a great measure, identified with the creoles, or Spanish Americans.

"The cholos (Peru and Bolivia)[14] come from a mixture of the mestizos and Indians: they are said to surpass all other classes, both in bodily strength and activity, and in native genius. They receive but little education, and speak indifferently the Indian and Spanish. [. . .][15]

"The proportion which these two classes bear to the others, it is almost impossible to ascertain: the gentleman from whom I received my estimate, assured me, that the inhabitants of pure blood, were not more than as one to fifteen; as many of those who are ranked as Spaniards, have a greater or less portion of Indian blood in their veins.

"The creoles constitute the third class in point of number. These again are divided into the nobles; such as counts, marquisses, mayorasgos (or owners of fiefs,) and knights of different military orders. These, of course, hold the most conspicuous rank in society; especially as they inherit large fortunes from their ancestors, the first conquerors and early adventurers.

"The eldest sons, who suceed [sic] to the estate, are, generally, but indifferently educated; and from the want of suitable objects to engage their minds, pass their time in idleness and dissipation. The number of these nobles, however, in the viceroyalty of La Plata, is inconsiderable, when compared to Lower Perú. The younger sons, if possessed of an inclination to study, become curates, or lawyers, explorers of mines, or they become owners of haciendas, or plantations, where they pursue the different cultures of the country.

"There are, usually, a number of families of Indians, whose duties are similar to those of the peasantry of Chile; the young Indians are their domestic servants. The higher clergy, as well as the monks and friars, in whose hands immense wealth has been amassed, are, in general, Europeans; but the secular clergy are Americans, and distinguished for their eloquence and learning. [. . .][16]

"The profession of the law forms a very numerous corps in these countries; and many of its members amass considerable fortunes. As legal proceedings are chiefly carried on in writing, their written eloquence surpasses their public speaking."[17]

We have drawn out this curious inventory of races, even pointing out their moral capacities and their occupations, because they should come to form part of the reader's understanding of the elements that make up our society, and of the influence that these casts and occupations should have over the new society that will be created when the Peninsular Spaniards, too, lose control of the position occupied by the tyrant class in the Quechua and Aztec empires.

This picture of the society is from the beginning of this century, and even if the racial proportions are not mathematically precise, they were accepted at the time, and the two generations that have passed since then is too short a time for them to have been significantly altered.

Now let us hear from wise Mr. Agassiz about the moral character of these races.

"Let any one who doubts the evil of this mixture of races, and is inclined, from a mistaken philanthropy, to break down all barriers between them, come to Brazil. He cannot deny the deterioration consequent upon an amalgamation of races, more widespread here than in any other country in the world, and which is rapidly effacing the best qualities of the white man, the negro, and the Indian, leaving a mongrel nondescript type, deficient in physical and mental energy" (Agassiz, p. 293).[18]

"The hybrid between White and Indian, called Mammeluco in Brazil, is pallid, effeminate, feeble, lazy, and rather obstinate; though it seems as if the Indian influence had only gone so far as to obliterate the higher characteristics of the White, without imparting its own energies to the offspring. It is very remarkable how, in both combinations, with Negroes as well as Whites, the Indian impresses his mark more deeply upon his progeny than the other races, and how readily, also, in further crossings, the pure Indian characteristics are reclaimed and those of the other races thrown off. I have known the offspring of an hybrid between Indian and Negro with an hybrid between Indian and White resume almost completely the characteristics of the pure Indian."[19]

"Another feature which makes a painful impression on the stranger is the enfeebled character of the population. I have spoken of this before [. . .].[20] It is not merely that the children are of every hue [. . .].[21] [With the crossing of three races][22] it is as if all clearness of type had been blurred, and the result is a vague compound lacking character and expression.

"This hybrid class, although more marked [in the North][23] because the Indian element is added, is very numerous in the cities and on the large plantation . . . "

THE BLACK RACE

Spanish America was, to be sure, the part of America that most needed a second servile race to save the natives from annihilation. A fact to make one meditate at length on the strange results of racial mixing is that the independence of the White race eliminated the Black race all across the continent. It lives in freedom in the United States alone, where it numbers around five million, following a social war. In Havana it remains enslaved, after a supreme and failed attempt by the White criollo race to emancipate itself. In Brazil the blacks that live today remain slaves, though their bondage is not passed on to their children.

The United States' war of secession was the fruit of an error in judgment. It was firmly believed that tropical crops could not be obtained at convenient prices without the work of that race whose cutis looks at though it had been charred by the rays of the sun. After the end of this war, once the rebels had been subdued and the blacks freed, cotton had to be grown again by any means necessary in order to live, with the work of free hands, with machines to save money. Ten years later the subjugated planters produced two and three times as many bales of cotton as before the war, and even at more lucrative prices. Ten million dollars and one million human lives had been lost, all for overlooking the fact that slavery, even as exploitation, is of no use in this day and age. Blacks now take part in U.S. politics like Indians in the rest of America; and perhaps the Whites there, at least in the South, in their children's generation, will have to atone for their ancestors' error. For they removed from Africa and their way of life, races that God has reserved for future worlds, perhaps for the one being prepared by Livingstone, Stanley, and Brazza, on the Congo and Zambezi Rivers and their tributaries.

And yet, nature itself, the secret and latent workings of affinity and repulsion, labors away in silence, without a plan and as if by instinct, until one day you look around and see neither the sons of the conquistadors, nor black slaves; the former on their way to disappearing, the latter wiped out in less than half a century all across Spanish America. In Chile there is not a one, in Lima there are but a few. In speaking of the blacks of México, Wilson says that he can only relate what he has heard, as he has not seen a single one. As for Buenos Aires, within twenty years one will have to go to Brazil in order to see them in their full racial purity.

However, in 1770, Buenos Aires had a total of 16,000 inhabitants, of whom one thousand were Spaniards from Europe; three or four thousand had been born in the country to Spanish parents, and were called criollos. All the rest of the inhabitants (eleven thousand) were Mulattoes, Half-Breeds, and Blacks.

"Blacks make up the largest number (1729), America is full of them, not because they have a separate nation, but rather because they are brought in from Africa, where they are bought from their parents. This is what they call the Slave Trade. The English carry them on their ships, and sell them at one and two hundred pesos a head. In all the cities subsumed under the generic name of Paraguay, the households use only black servants, because no Spaniard, poor as he may be, wants to take up servanthood. As far as the Indians are concerned, few are to be seen in the Spanish cities; and those that one does see are the ones who can come and go freely. It is very rare for them to enter into the service of the Spanish; and they no longer dare to attack their liberty as before. The Spanish have had more than one opportunity to regret their erstwhile violence."[24]

Buenos Aires owes some service to the Blacks, who helped beautify it. "Buenos Aires," says the same Jesuit traveler, "holds, without question, the first place among all the cities that the Spanish have built from the Mountains to the Ocean, including Asunción with its ten thousand inhabitants.

"As far as Buenos Aires is concerned, though one sees, as in the other cities, houses randomly strewn about and surrounded by trees, they make up more or less straight and tidy streets. It is true that the oldest of these houses are made of clay and are only one story high. This is because it was only recently that one of our brothers, who had been sent here from Europe to build our church, figured out how to fire bricks in this country, which today boasts more than sixty kilns. Now one can see some two-story houses . . . Another considerable service that the Spanish have done is to make a large number of craftsmen out of their Black servants. Now one need only show them a design and they carry it out perfectly. In this way Buenos Aires is made more and more beautiful by the day, and quite soon it will have something to offer to the eyes of Europeans" (1726).

Though this is account is very old, I can recall having met, in my own time in San Juan, the master craftsman Antonio, who was a slave of the house of Don Pedro Vásquez del Carril. He was a very respectable man, respected by one and all, much like Alejos, the renowned master blacksmith, who was a great figure in politics and one of the most skilled artisans in his profession.

In 1810 this American world began to stir, and break apart into pieces, like the thawing of a great river whose waters had long since been held back, a river of races and social elements that were only held together by a lack of heat.

The war of independence required firm spirits; and it has already been seen how the Indians sided more with the conquistadors, or else were indifferent. After the battle of Salta, Belgrano surrendered 3,000 of them. He freed them on their own word of honor, and still they took up arms again, because they did not know the meaning of honor, and because once again the Spanish were in need of them.

Though he was a slave, the Negro was a friend of his young criollo master, perhaps having been raised alongside him in the same family, having played the same games and acquired the same tastes. By his race he is faithful and enthusiastic, and so he volunteered to serve as an assistant and accompany his "master" in the war.

Why not set up battalions then, and have their masters either give them their freedom or offer them to their country as their own contribution of blood? Blacks made up numbers 7 and 8, famous in the war in Chile and Perú; number 9 and number 10, which formed part of the forces at Desaguadero; number 2, which came back from Brazil. Yet another company of these brave veterans, with their black faces and white heads, died in the lagoons of Hanacache in 1831, alongside Commander Castro, caught unawares by Quiroga's forces.

In 1826, with all of the Black African slaves that were taken prisoner, the war in Brazil ended up renewing the race's stock in Buenos Aires, after it had been decimated by so much bloodshed. The attempts to settle them on the ranches were not very successful due to the cold, which is why there was a great number of them in the city.

Each African nation, the Guineans, the Mandingas, the Congos, set up their own municipal establishments called *Candombes*, on account of the drum used to accompany their dancing, which for the African is the expression of life and happiness. A king and queen are named to preside over the festival, keep order, and collect contributions and alms to bury the dead in their parish and to help the sick.

The *Candombes* were the terror of Buenos Aires during the tyranny of Rosas, who made Manuelita into the patroness of the institution. One day they strolled through the streets of Buenos Aires, drunk with enthusiasm, with their *candombes* and marimbas ahead of them, those Africans grouped together in patriotic clubs, behind red flags, like today in the societies of France, Spain, and Italy, and a music group at the front. A day of horror for the Whites, for the sons of the Spanish, who prepared, executed, and carried out the Independence. They were now banished and handed over to hellish gods, to cries of: "Let the savage Centralists die! Long live the great Restorer!" that one thousand glistening black mouths called out.

The Black race also gave its contribution of blood to the war of extermination.

In Montevideo four battalions were raised from among the young black men enclosed in the besieged city, while Rosas sent out a regiment of Blacks in order to tighten the siege, which had been reduced to thirty soldiers sent out by a black sergeant when the author of *Ejército Grande*[25] came upon in 1851, the sole survivors of that Trojan siege.

In Buenos Aires the man who had won the battle at Caseros rounded up as many colored men as he could and sent them on to Calá, in Entre Ríos Province, to serve as members of the infantry with which he intended to fortify his squadrons of lancers. Few of them made it back home.

Dark-skinned men traditionally made up divisions of the urban militia, alongside the patricians. After the separation of September 11th, 1852 there were two large battalions of 1,800 men commanded by colored officers, such as Colonel Sosa and later on Colonel Morales, because the highest and the lowest ranks of the races were always brought together by links of reciprocal fondness.

In the Chambers the latter was likewise represented by Sosa and Mendizábal, and the most able members of their stock. They were in attendance at Cepeda, were present in Pavón, and were even painfully represented in Los Corrales.

There are just a few young people of color left, and they take up service as top coachmen, as doormen at public offices, and other lucrative jobs. Yet as a race, as a social element, they are now strictly incidental, since they have altogether disappeared in the provinces and have not been able to establish themselves anywhere outside the city.

As industrialists they have not built fortunes, a phenomenon that can be seen in the United States, despite the fact that in the banks they have deposited millions in savings.

We shall end this sad chapter in the forced removal of human races with this prophesy that was spoken by the heart, and is now being fulfilled by history.

"If ever Africa shall show an elevated and cultivated race—and come it must, some time, her turn to figure in the great drama of human improvement—life will awake there with a gorgeousness and splendour of which our cold western tribes faintly have conceived. In that far-off mystic land of gold and gems, and spices, and waving palms, and wondrous flowers, and miraculous fertility, will awake new forms of art, new styles of splendour; and the negro race, no longer despised and trodden down, will, perhaps, show forth some of the latest and most magnificent revelations of human life. Certainly they will, in their gentleness, their lowly docility of heart, their aptitude to repose on a superior mind and rest on a higher power, their childlike simplicity of affection, and facility of forgiveness. In all these they will exhibit the highest form of the peculiarly *Christian life*, and, perhaps, as God chasteneth whom he loveth, he hath chosen poor Africa in the furnace of affliction, to make her the highest and noblest in that kingdom which he will set up when every other kingdom has been tried and failed; for the first shall be the last, and the last first."[26]

And this prophesy, inspired by the intuitions of woman's motherly love, is now on its way to being fulfilled, with a splendor, a solidarity, and a momentum that has left behind the sixteenth century, when only kings took up the conquest of America, and that lends it the air of a truly providential event. Just as, in the wake of Columbus's caravels, ships from every nation raced to capture their share of the spoils, in the same way, behind the slow but firm steps of the Christian priest Livingstone—the Apostle Paul of the Black race, who holds up Christian virtues as the only way to seduce the Blacks—all the heroes and greats of modern thought have followed. We have Stanley, the heroic reporter from the *Herald*, the foremost daily in the United States, the representatives of Italy, of Prussia in other directions, France extending down into the South from its African possessions on new railroads, and even England, in the White, or Felata, or Arab Africa of the Mediterranean, as in the extreme South, with Cetewayo, and the eastern banks of the Zambezi, and the Diamond mines. The world is full of reports of Africa alone, of the discoveries, magnificence, and splendors of Africa, because everyone feels that its time for justice, for dignity, and for redress has finally come.

NOTES

1. OAN: New History of the Conquest of Mexico—Wilson. [EN: Wilson, Robert Anderson. *A New History of the Conquest of Mexico, in Which Las Casas' Denunciations of the Popular Historians of That War Are Fully Vindicated*. Philadelphia: James Challen and Sons, 1859.]

2. TN: Sarmiento omits the following: "It will now be easily understood why it was that, in all important respects, the civilizations of Mexico and Peru were strictly analogous to those of India and Egypt. In these four countries, as well as in a few others in Southern Asia and Central America, there existed an amount of knowledge, despicable indeed if tried by an European standard, but most remarkable if contrasted with the gross ignorance which prevailed amongst the adjoining and cotemporary [sic] nations. But in all of them . . . "

3. EN: Henry Thomas Buckle, *History of Civilization in England*, vol. 1 (London: Longmans, Green, and Co, 1868), 112—This text is for some reason cited by Sarmiento as part of Prescott, 1862, though Buckle is elsewhere mentioned in this chapter.

4. EN: Ibid, 114.

5. EN: William Hickling Prescott, *History of the Conquest of Peru, with a Preliminary View of the Civilization of the Incas*, vol. 1 (London: Routledge, Warne, and Routledge, 1862), 111.

6. OAN: *Voyages*—F. Depons, book 1, pages 238 and 239.

7. EN: Wilson, 33–34.

8. TN: Sarmiento omits the following from the original Wilson text (ellipsis mine): "He saw only the cruelty the Indians suffered, and the rapid progress of their decay; and concluded the one to be the cause of the other, without suspecting, that a higher law was involved in the result; a law whose operation neither cruelty nor kindness could materially influence."

9. OAN: Wilson, *A New History of Mexico*, 35.

10. TN: *mestizos*.

11. OAN: Blackenridge [sic], p. 81. [EN: Brackenridge, H. M. *Voyage to South America, Performed by Order of the American Government, in the Years 1817 and 1818, in the Frigate Congress*, vol. 2. (Boston: published by the author, 1819), 81].

12. EN: Ibid., 111.

13. EN: Ibid., 81.

14. TN: This parenthetical aside belongs to Sarmiento, not Brackenridge.

15. TN: Sarmiento omits a short passage in which Brackenridge notes how the "cholos" are "said to surpass all other classes, both in bodily strength and activity, and in native genius," as well as being "admirable soldiers," and the "principle leaders in the present revolution." Ellipsis mine.

16. TN: Omits a short passage about the secular clergy's involvement in law and also in the revolution, and their "disposition to throw off the papal yoke."

17. OAN: ibid, 85–87.

18. TN: This citation belongs to Sarmiento.

19. EN: Mr and Mrs Louis Agassiz, *A Journey in Brazil* (Boston: Ticknor & Fields, 1863), 532.

20. TN: Omits the following from the original Agassiz text (ellipsis mine): "but in the northern provinces it is more evident than farther south."

21. TN: Omits the following from the original Agassiz text (ellipsis mine): "the variety of color in every society where slavery prevails tells the same story of

amalgamation of race; but here this mixture of races seems to have had a much more unfavorable influence on the physical development than in the United States."

22. TN: Added by Sarmiento.

23. TN: Added by Sarmiento.

24. OAN: Lettre Seconde, du Père Gaetan Muratori, du Paraguay.

25. TN: *Big Army*.

26. OAN: *Uncle Tom's Cabin*; by Mrs. Beecher Stowe, a book that has been translated into all of the modern civilized languages.

BIBLIOGRAPHY

Alberdi, Juan Bautista. *Escritos póstumos: Ensayos.* Buenos Aires: Imprenta Europea, 1895–1901.

Ameghino, Florentino. *Doctrinas y descubrimientos.* Buenos Aires: La Cultura Argentina, 1917.

——— "Prólogo." In *Escritos científicos* by Francisco Javier Muñiz, xiii–xviii. Buenos Aires: W. M. Jackson, 1953.

Ameghino, Florentino, and Carlos Ameghino. *Reseñas de la Patagonia: Andanzas, penurias y descubrimientos de dos pioneros de la ciencia.* Buenos Aires: Ediciones Continente, 2006.

Argüello, Isaura. *La actuación internacional de Francisco P. Moreno.* Buenos Aires: Universidad de Buenos Aires. Facultad de Ciencias Económicas, 1952.

Asúa, Miguel. *Ciencia y literatura: Un relato histórico.* Buenos Aires: Eudeba, 2004.

———, ed. *La ciencia en la Argentina: Perspectivas históricas.* Buenos Aires: CEAL, 1993.

Azara, Félix de. *Apuntamientos para la historia natural de los páxaros del Paraguay y del Río de la Plata.* Madrid: Imprenta de la Viuda de Ibarra, 1802.

Babini, José. "Breve historia de la ciencia argentina." In *La ciencia en la Argentina: Perspectivas históricas*, edited by Miguel Asúa, 27–43. Buenos Aires: CEAL, 1993.

Barnett, S. A. *A Century of Darwin.* London: Heinemann Educational, 1962.

Beer, Gillian. *Darwin's Plots: Evolutionary Narrative in Darwin, George Eliot, and Nineteenth Century Fiction.* London: Routledge, 1983.

Belluzo, Ana Maria de Moraes. *Un lugar no Universo: O Brasil dos Viajantes*, vol. 2. Sao Paulo: Metalivros, 1994.

Bertomeu, Carlos. *El perito Moreno, centinela de la Patagonia.* Buenos Aires: El Ateneo, 1949.

Bernasconi, Robert, and Tommy Lott. *The Idea of Race.* Indianapolis: Hackett, 2000.

Biagini, Hugo, ed. *El movimiento positivista argentino.* Buenos Aires: Editorial Belgrano, 1985.

Bourdieu, Pierre. *The Rules of Art.* Stanford: Stanford University Press, 1996

Brahn, Paul. *The Cambridge Illustrated History of Archaeology*. New York: Cambridge University Press, 1996.

Brown, Bill. *Things*. Chicago: University of Chicago Press, 2004.

Carlson, Elof Axel. *The Unfit: A History of a Bad Idea*. Cold Spring Harbor: Cold Spring Harbor Laboratory Press, 2001.

Cayley, David. *Ideas on the Nature of Science*. Fredericton: Goose Lane Editions, 2009.

Clarke, Angus, and Flo Ticehurst. *Living with the Genome: Ethical and Social Aspects of Human Genetics*. New York: Palgrave Macmillan, 2006.

Clifford, James. *The Predicament of Culture*. Boston: Harvard University Press, 1988.

Cohen, Samuel. *50 Essays: A Portable Anthology*. Boston: Bedford/St. Martin's, 2007.

Conkin, Paul. *When All the Gods Trembled: Darwinism, Scopes, and American Intellectuals*. Lanham: Rowman & Littlefield, 1998.

Crook, D. P. *Darwin's Coat-Tails: Essays on Social Darwinism*. New York: Lang, 2007.

Dalziel, Ian. *In Darwin's Footsteps: The Geology of Tierra del Fuego*. London: Geological Society of London, 2007.

Darwin, Charles. *From So Simple a Beginning: The Four Great Books of Charles Darwin*. Edited by Edward Wilson. New York: Norton, 2006.

Degiovanni, Fernando. *Los textos de la patria*. Rosario: Beatriz Viterbo, 2007.

De Paolo, Charles. *The Ethnography of Charles Darwin: A Study of His Writings on Aboriginal Peoples*. Jefferson: McFarland, 2010.

Donghi, Halperín, et al. *Sarmiento: Author of a Nation*. Berkeley: University of California Press, 1994.

Ede, Andrew, and Lesley Cormack. *A History of Science in Society*. Peterborough: Broadview, 2004.

Fabbri, Marta. "Inserciones de lo fantástico en las *Fuerzas extrañas*." In *El relato fantástico rioplatense en el siglo XIX*, edited by Enriqueta Morillas Ventura, 53–73. Neuquén: Universidad Nacional del Comahue, 2005.

Fasano, Héctor. *Perito Francisco Pascasio Moreno: Un héroe civil*. La Plata: Universidad Católica de La Plata, 2002.

Fernández Bravo, Álvaro. "Latinoamericanismo y representación: Iconografías de la nacionalidad en las exposiciones universales (París 1889–1900)." In *La ciencia en la Argentina entre siglos*, edited by Marcelo Montserrat, 171–85. Buenos Aires: Manantial, 2000.

Fernández, Laura. "La pampa de memoria: William Henry Hudson." *Ciberletras* 9 (2003). http://www.lehman.cuny.edu/ciberletras/v09/fernandez.html.

Fishburn, Evelyn. *The Portrayal of Immigration in Nineteenth Century Argentine Fiction (1845–1902)*. Berlin: Colloquium Verlag, 1981.

Ford, Aníbal. "Darwin, Fitz-Roy y los intereses ingleses en el Atlántico Sur: Ensayo de contrainformación cultural." In *Desde la orilla de la ciencia: Ensayos sobre identidad, cultura y territorio*, edited by Aníbal Ford, 67–82. Buenos Aires: Punto Sur, 1988.

Franco, Jean. "Un viaje poco romántico." *Escritura* 7 (Jan–June 1979): 129–42.

Franco, Luis. "Hudson en la Pampa." *Revista hispánica Moderna* (Jan–Apr 1944): 99–106.

Frederick, John T. *William Henry Hudson*. New York: Twayne, 1972.

Fung de Lanning, Rosa. "Las ideas evolucionistas en las interpretaciones arqueológicas." *Revista del Museo Nacional* 32 (1963): 203–28.

Gallo, Antonio. "Hudson y lo nacional." *Revista Hispánica Moderna* (July–Oct 1944): 308–11.

García, Juan Agustín. *Introducción al estudio de las ciencias sociales argentinas*. Buenos Aires: Imprenta Coni, 1899.

Gasparini, Sandra. "Introducción." In *Dos partidos en lucha*, by Eduardo Holmberg, 9–39. Buenos Aires: Corregidor, 2005.

Glick, Thomas, Miguel Angel Puig-Samper, and Rosaura Ruiz, eds. *El Darwinismo en España e Iberoamérica*. México: Universidad Autónoma de México: 1999.

González-Echeverría, Roberto. *Myth and Archive*. New York: Cambridge University Press, 1990.

González-Stephan, Beatriz. "Showcases of Consumption: Historical Panoramas and Universal Expositions." In *Beyond Imagined Communities*, edited by Sara Castro and John Chasteen, 225–38. Baltimore: Johns Hopkins University Press, 2003.

González-Stephan, Beatriz, and Jens Andermann. *Galerías del progreso: Museos, exhibiciones y cultura visual en América Latina*. Rosario: Beatriz Viterbo, 2006.

Graves, Joseph L. *The Emperor's New Clothes: Biological Theories of Race at the Millennium*. New Brunswick: Rutgers University Press, 2001.

Gravil, Roger. *The Anglo-Argentine Connection, 1900–1939*. Boulder and London: Westview, 1985.

Hanks, Craig. *Technology and Values: Essential Readings*. Chichester and Malden: Wiley-Blackwell, 2010.

Hemleben, Joannes. *Darwin*. Madrid: Alianza, 1971.

Hicken, Cristóbal. "Eduardo. L. Holmberg y las doctrinas evolucionistas." In *El joven coleccionista de historia natural en la República Argentina*, by Eduardo Holmberg, 13–28. Buenos Aires: Publicaciones de la Sociedad Luz, 1905.

Hoeg, Jerry, and Kevin S. Larsen, eds. *Interdisciplinary essays on Darwinism in Hispanic Literature and Film: The Intersection of Science and the Humanities*. Lewiston: Mellen, 2009.

Holmberg, Eduardo Ladislao. *Dos partidos en lucha*, edited by Sandra Gasparini. Buenos Aires: Corregidor, 2005.

———. *El joven coleccionista de historia natural en la República Argentina*. Buenos Aires: Publicaciones de la Sociedad Luz, 1905.

Hudson, William Henry. *Birds of La Plata*. London and New York: Dent & Sons/ Dutton, 1920.

———. *Far Away and Long Ago*. New York: Dutton, 1918.

———. *A Naturalist in La Plata*. New York: Dover, 1988 [1892].

Huertas, Rafael. *El delincuente y su patología: Medicina, crimen y sociedad en el positivismo Argentino*. Madrid: Consejo Superior de Investigaciones Científicas, 1991.

Ingenieros, José. *Las doctrinas de Ameghino: La tierra, la vida y el hombre*. Buenos Aires: Talleres de J. Rosso y Cía., 1919.

———. *Sociología argentina*. Madrid: Daniel Jorro, 1913.

Jackson, John, and Nadine Weidman. *Race, Racism, and Science: Social Impact and Interaction*. Santa Barbara: ABC-CLIO, 2004.

Jagoe, Eva-Lynn Alicia. *The End of the World as They Knew It: Writing Experiences of the Argentine South*. Lewisburg: Bucknell University Press, 2007.

Jensen, J. Vernon. *Thomas Henry Huxley: Communicating for Science*. Newark: University of Delaware, 1991.

Kitcher, Philip. *In Mendel's mirror: Philosophical Reflections on Biology*. Oxford and New York: Oxford University Press, 2003.

Latour, Bruno. *Science in Action: How to Follow Scientists and Engineers through Society*. Milton Keynes: Open University Press, 1987.

Leask, Nigel. *Curiosity and Aesthetics of Travel Writing*. New York: Oxford University Press, 2002.

Lewens, Tim. *Darwin*. London: Routledge, 2007.

Livón-Grosman, Ernesto. *Geografías imaginarias: El relato de viaje y la construcción del espacio patagónico*. Rosario: Beatriz Viterbo, 2003.

Lugones, Leopoldo. *Elogio de Ameghino*. Buenos Aires: Otero, 1915.

Mac Donagh, Emiliano. "La ciencia argentina en la vida de Hudson." *La Nación*, May 23, 1943, sec. Cultural.

———. *150 años de evolución científica argentino-británica*. Buenos Aires: Comisión de Investigación Científica, 1960.

Mantegari, Cristina. *Germán Burmeister: La institucionalización científica en la Argentina del siglo XIX*. Buenos Aires: Jorge Baudino Ediciones/UNSAM, 2003.

Mañé Garzón, Fernando. *Un siglo de darwinismo: Un ensayo sobre la historia del pensamiento biológico en Uruguay*. Montevideo: Facultad de Medicina, 1990.

Mayo, Carlos. *El positivismo en la política argentina 1880/1906*. Buenos Aires: CEAL, 1988.

Mignolo, Walter. *The Dark Side of the Renaissance*. Ann Arbor: University of Michigan Press, 1995.

Miranda, Marisa, and Gustavo Vallejo, eds. *Darwinismo social y eugenesia en el mundo latino*. Buenos Aires: Siglo XXI, 2005.

Montserrat, Marcelo. "La mentalidad evolucionista: Una ideología del progreso." In *Ciencia, historia y sociedad en la Argentina del siglo XIX*, edited by Marcelo Monserrat, 31–69. Buenos Aires: CEAL, 1993.

———. "La presencia evolucionista en el positivismo argentino." In *Ciencia, historia y sociedad en la Argentina del siglo XIX*, edited by Marcelo Monserrat, 70–82. Buenos Aires: CEAL, 1993.

———. "La recepción literaria de la ciencia en la Argentina: El caso darwiniano." In *Usos de la memoria: Razón, ideología e imaginación históricas*, edited by Marcelo Montserrat, 150–69. Buenos Aires: Sudamericana, 1996.

———. "La sensibilidad evolucionista en la Argentina decimonónica." In *La ciencia en Argentina entre siglos*, edited by Marcelo Monserrat et al., 203–21. Buenos Aires: Manantial, 2000.

Moorehead, Alan. *Darwin y el Beagle (1831–1836)*. Barcelona: Ediciones del Aguazul, 2002.

Moreno, Francisco. *Reminiscencias del Perito Moreno*. Buenos Aires: El Elefante Blanco, 1997.

———. *Viaje a la Patagonia Austral, emprendido bajo los auspicios del gobierno nacional, 1876–1877*. Buenos Aires: Imprenta de La Nación, 1879.

Nouzeilles, Gabriela. *Ficciones somáticas*. Rosario: Beatriz Viterbo, 2000.

Novoa, Adriana. "The Dilemmas of Male Consumption in Nineteenth-Century Argentina: Fashion, Consumerism, and Darwinism in Domingo Sarmiento and Juan B. Alberdi." *Journal of Latin American Studies* 39 (2007): 771–96.

Novoa, Adriana, and Alex Levine. *From Man to Ape: Darwinism in Argentina 1870–1920*. Chicago: University of Chicago Press, 2010.

Orioni, Julio, and Fernando Rocchi. "El darwinismo en la Argentina." *Todo es historia* 228 (1986): 8–26.

Pagden, Anthony. *European Encounters with the New World: From Renaissance to Romanticism*. New Haven and London: Yale University Press, 1993.

Parsons, Keith M. *The Science Wars: Debating Scientific Knowledge and Technology*. Amherst: Prometheus, 2003.

Pennini de la Vega, Elena. "Darwin en Argentina." *Quipus* 1 (1984): 119–32.

Pichot, André. *The Pure Society: From Darwin to Hitler*. London and New York: Verso, 2009.

Podgorni, Irina. "De la santidad laica del científico Florentino Ameghino y el espectáculo de la ciencia en la Argentina Moderna." *Entrepasados: Revista de Historia* 6, no. 13 (1997): 37–61.

———. "Los gliptodontes en París: Las colecciones de mamíferos fósiles pampeanos en los museos europeos del siglo XIX." In *La ciencia en la Argentina entre siglos*, edited by Marcelo Montserrat, 309–27. Buenos Aires: Manantial, 2000.

———. "Prólogo." In *Patagonia: Ciencia y conquista; La mirada de la primera comunidad científica argentina*, edited by Pedro Navarro Floria, 9–29. Neuquén: CEP, Universidad Nacional del Comahue, 2004.

———. *El sendero del tiempo y de las causas accidentales: Los espacios de la prehistoria en la Argentina, 1850–1910*. Rosario: Prohistoria, 2009.

Pratt, Mary Louise. *Imperial Eyes: Travel Writing and Transculturation*. London: Routledge, 1992.

Prieto, Adolfo. *Los viajeros ingleses y la emergencia de la literatura argentina (1820–1850)*. Buenos Aires: Sudamericana, 1996.

Prodger, Phillip. *Darwin's Camera: Art and Photography in the Theory of Evolution*. Oxford and New York: Oxford University Press, 2009.

Ramos Mejía, Francisco. *Historia de la evolución argentina*. Buenos Aires: Librería "La Facultad" de Juan Roldán, 1921.

Richards, Robert. *Darwin and the Emergence of Evolutionary Theories of Mind and Behavior.* Chicago: University of Chicago Press, 1987.

Rodríguez, Fermín. "Prehistorias argentinas: Naturalistas en el Plata; Charles Darwin, Francisco Moreno, Florentino Ameghino, Bruce Chatwin." *A Contracorriente* 7, no. 1 (2009): 45–75.

Rojas, Ricardo. *Los arquetipos*. Buenos Aires: Roldán, 1922.

Romano, Eduardo. "Fray Mocho: El costumbrismo hacia 1900." *Historia de la literatura argentina*, vol. 2, 265–279. Buenos Aires: CEAL, 1980.

Romero, José Luis. *Las ideas políticas en Argentina*. Buenos Aires: FCE, 1959.

Ruiz Gutiérrez, Rosaura, and Francisco Ayala. *De Darwin al DNA*. México: Fondo de Cultura Económica, 2002.

Ruse, Michael. *Charles Darwin*. Oxford: Blackwell, 2008.

Salgado, Leonardo, and Pedro Navarro Floria. "Hermann Burmeister y su *Historia de la creación*: Idealismo, materialismo y empirismo en el credo de la primera ciencia argentina." In *Patagonia: Ciencia y conquista; La mirada de la primera comunidad científica argentina*, edited by Pedro Navarro Floria, 37–60. Neuquén: CEP, Universidad Nacional del Comahue, 2004.

Said, Edward. *Orientalism*. New York: Pantheon, 1978.

Salessi, Jorge. *Médicos, maleantes y maricas*. Rosario: Beatriz Viterbo, 1995.

Salvatore, Ricardo, comp. *Culturas imperiales.* Rosario, Beatriz Viterbo, 2005.

Sarlo, Beatriz. *Una modernidad periférica*. Buenos Aires: Nueva Visión, 1988.

Sarmiento, Domingo F. "Mundos prehistóricos: Viaje aéreo a través del Museo Prehistórico de Moreno." In *Obras*, 135–45. Buenos Aires: Imprenta Mariano Moreno, 1899.

———. *Viajes por Europa, África y América*. Madrid: Fondo de Cultura Económica, 1996.

———. *Vida y escritos del coronel D. Francisco J. Muñiz*. Buenos Aires: Félix Lajouane, 1885.

Schmitt, Cannon. *Darwin and the Memory of the Human: Evolution, Savages, and South America*. New York: Cambridge University Press, 2009.

Sewell, Dennis. *The Political Gene: How Darwin's Ideas Changed Politics*. London: Picador, 2009.

Shanahan, Timothy. *The Evolution of Darwinism*. Cambridge: Cambridge University Press, 2004.

Soler, Ricaurte. *El positivismo argentino*. Panamá: Imprenta Nacional, 1959.

Spencer, Mark. *Race, Gender, and Supremacy*. Bristol: Thoemmes, 2001.

Spurr, David. *The Rethoric of Empire.* Durham: Duke University Press, 1993.

Stockwell, Sarah. *The British Empire: Themes and Perspectives.* Oxford: Blackwell, 2008.

Terán, Oscar. *Positivismo y nación en la Argentina*. Buenos Aires: Puntosur, 1987.

Teruggi, Mario. *Museo de La Plata, 1888–1988: Una centuria de honra.* La Plata: Fundación Museo de La Plata, 1994.

Vandenberghe, Frédéric. *Complexités du posthumanisme: Trois essais dialectiques sur la sociologie de Bruno Latour.* Paris: L'Harmattan, 2006.

Veyga, Francisco de. "De la regeneración como ley opuesta a la degeneración mórbida." *Archivos de psiquiatría y criminología* 4 (1905): 31–44.

Ygobone, Aquiles. *Francisco P. Moreno: Arquetipo de argentinidad.* Buenos Aires: Plus Ultra, 1979.

Weimberg, Gregorio. *La ciencia y la idea del progreso en América Latina, 1860–1930.* México: Fondo de Cultura Económica, 1996.

Zuretti, Juan Carlos. *Historia de la cultura argentina: Arte, ciencia.* Buenos Aires: Itinerarium, 1964.

INDEX

ABOUT THE EDITOR

Leila Gómez is associate professor of Spanish at the University of Colorado at Boulder.

CPSIA information can be obtained at www.ICGtesting.com
Printed in the USA
BVOW012343131111

275631BV00003BB/1/P

9 781611 483864